# THE GLOBAL NUCLEAR LANDSCAPE

## Energy, Non-proliferation and Disarmament

# THE GLOBAL NUCLEAR LANDSCAPE

## Energy, Non-proliferation and Disarmament

Edited by

**Manpreet Sethi**

Introduction by

Air Marshal **Anil Chopra** PVSM AVSM VM VSM (Retd)

Centre for Air Power Studies
New Delhi

*in association with*

KNOWLEDGE WORLD
KW Publishers Pvt Ltd
New Delhi

The Centre for Air Power Studies is an independent, non-profit, academic research institution established in 2002 under a registered Trust to undertake and promote policy-related research, study and discussion on the trends and developments in defence and military issues, especially air power and the aerospace arena, for civil and military purposes. Its publications seek to expand and deepen the understanding of defence, military power, air power and aerospace issues without necessarily reflecting the views of any institution or individuals except those of the authors.

Centre for Air Power Studies
P-284, Arjan Path
Subroto Park
New Delhi 110010

Tele: (91-11) 25699131

E-mail: capsnetdroff@gmail.com
website: www.capsindia.org



ISBN 978-93-91490-53-9 Hardback
ISBN 978-93-91490-57-7 ebook

Published in India by Kalpana Shukla

KNOWLEDGE WORLD

KW Publishers Pvt Ltd
4676/21, First Floor, Ansari Road
Daryaganj, New Delhi 110002
Phone: +91 11 43528107
Marketing: kw@kwpub.in
Editorial: production@kwpub.in
Website: www.kwpub.in

Printed and bound in India

## CENTRE FOR AIR POWER STUDIES

### VISION

To be an independent centre of excellence on national security contributing informed and considered research and analyses on relevant issues.

### MISSION

To encourage independent and informed research and analyses on issues of relevance to national security and to create a pool of domain experts to provide considered inputs to decision-makers. Also, to foster informed public debate and opinion on relevant issues and to engage with other think-tanks and stakeholders within India and abroad to provide an Indian perspective.

# Contents

# Preface

Like the shifting sands of a desert, the global nuclear landscape too changes every few years across its three main constituents: nuclear energy, non-proliferation and disarmament. Each of them has been visited by phases of cautious optimism, deep scepticism and outright pessimism over just the last two decades. This book offers a bird's eye view on all the three, even as the individual authors offer a worm's eye view on each specific topic within the larger ambit.

## Nuclear Energy

The millennium started with the promise of a new renaissance for nuclear energy. Sixteen percent of the total global electricity generation was through nuclear energy in 2000; 31 reactors were under construction even as 438 were operational across the world. At the end of 2009, the International Atomic Energy Agency (IAEA) had identified as many as 65 countries that had expressed an interest in starting or expanding national nuclear power programmes. However, the unfortunate accident at Fukushima in March 2011 changed the mood. The nuclear aspirants decided to put their plans on hold. Several of those already invested in nuclear power took a pause, conducted elaborate safety reviews, and moderated their ambitions for the future. A few announced a phased withdrawal of their programme. With public perception of safety around nuclear energy taking a nosedive, the second decade of the 2000s started with a shroud of scepticism over this sector.

A decade since Fukushima, the winds seem to have shifted the sands on the nuclear energy landscape. With more stringent safety designs, mechanisms, processes and regulations in place, many countries have begun to reconsider the desirability and feasibility of nuclear energy. Moreover, this change appears to be driven by the harsh reality of climate change. As more and more severe climatic events visit mankind with greater frequency, the need for urgent action by all nations to arrest their greenhouse gas emissions into the atmosphere is being recognised.

The Paris Agreement of 2015 envisions a global response to the threat of climate change and urges nations to arrive at their own strategies to contribute to keeping the global temperature rise this century to below 2 degrees Celsius above pre-industrial levels, and ideally, to limit the temperature increase to below 1.5 degrees Celsius. In the developing countries that are desirous of fulfilling this commitment, but that can also not compromise on their economic growth and development, of which electricity is an important constituent, the role of nuclear energy as a low carbon, baseload source of electricity is being ever more recognised. Meanwhile, optimism in this sector is also visible in the research and development on new technologies, such as small modular reactors and floating nuclear power plants.

The first section of this book is devoted to an assessment of developments taking place in the nuclear energy sector. An introductory chapter provides a broad-brush scan of the global picture, including contemporary drivers and challenges for nuclear energy. The second chapter, then, follows a similar approach to examine the particular situation of India. The following two chapters undertake an assessment of the concept, developments and prospects of small modular reactors. These are examined for both civilian and military applications. As it emerges, there is a case for cautious optimism as these technologies offer the promise of many advantages. But their fulfillment will be determined by how effectively and quickly the issues related to licensing and regulation are resolved, enabling the new concepts to prove their commercial viability.

## Nuclear Non-Proliferation

On this front, the landscape shifts have been quite dramatic. If the Review Conference (RevCon) of the Treaty on Non-Proliferation of Nuclear Weapons (NPT) can be taken as a guide on the state of non-proliferation, then the RevCon of 2000 generated a feel-good mood. For the first time in three decades since the treaty's entry into force in 1970, the states-parties were able to adopt, by consensus, a fully negotiated final document; 187 members of the NPT agreed on 13 practical steps for further progress on nuclear non-proliferation, including on safeguards, export controls and a strengthened review process, besides nuclear disarmament!

Soon after this RevCon though, the non-proliferation picture began to change. In October 2003, the German ship *BBC China* was seized and found to be carrying parts for a clandestine nuclear weapons programme of Libya. Details of an elaborate proliferation network surfaced from this incident. Pakistan emerged at the centre of this enterprise. AQ Khan was made to individually own up to these "unauthorised proliferation activities" in a televised disclosure on February 4, 2004. However, in an interview in 2011 to *Der Spiegel*, he stated, "I took sole blame for this whole episode because the political leadership urgently asked me to do so...". Pakistan wiggled its way out, but the network revealed the country's support for the clandestine weapons related efforts of Iraq, Iran, Libya, and North Korea. Of these, North Korea exited the NPT in January 2003.

As a consequence of these developments, the non-proliferation regime appeared to be in peril by the mid-2000s. In an effort to address the situation, the IAEA tightened its safeguards; nations reinforced their domestic export controls with updated and harmonised trigger lists; the UN Security Council passed Resolution 1540 mandating national action on criminalisation of unauthorised activities around Weapons of Mass Destruction (WMD) and their delivery systems. Amidst all this, North Korea conducted its first nuclear test in 2006. It had conducted five more by 2017. At the time of writing this piece, the possibility of another nuclear test by Pyongyang is in the air. Despite being under heavy sanctions, it has continued to advance its nuclear and missile capabilities. A resolution to the knotty North Korean problem is not in sight, thirty years after it first surfaced.

Meanwhile, one important concern that the international community did manage to successfully address in the decade of the 2010s was that of proliferation of nuclear material, technology and equipment to non-state actors. This was facilitated by the personal initiative of then US President, Barack Obama, who prioritised the threat of nuclear terrorism. By convening a series of four Nuclear Security Summits, he commandeered the highest level of political oversight to nuclear security. National leaders came together to accept domestic and international actions that would minimise the chances of nuclear and radiological material falling prey to the malicious intent of non-state actors.

Another proliferation concern that generated optimism around its possible resolution, albeit only momentarily, was that of Iran's nuclear programme. Long and arduous negotiations yielded the Joint Comprehensive Plan of Action (JCPOA) in 2015. However, in less than three years of the start of implementation of the agreement, Donald Trump, then US president decided to withdraw from the agreement. The JCPOA has since moved into a delicate phase since Iran too has been taking steps out of the agreement from 2019 onwards. Negotiations to return to the deal have not been successful. The picture looks bleak as we go into print in June 2022.

Meanwhile, another big blow to non-proliferation may have been dealt this year by the ongoing Russia-Ukraine conflict. The manner in which the use of force has been undertaken by a Nuclear Weapon State (NWS) against a Non-Nuclear Weapon State (NNWS) has enhanced the perception of utility of nuclear weapons. This could draw the NNWS towards nuclear weapons, especially if they face a hostile relationship with another nuclear armed country. The possibility of denuclearisation of North Korea has receded. Iran could be looking at nuclear weapons with new eyes, as some seem to be doing also in Japan and South Korea. The coming years are likely to be turbulent for non-proliferation. It appears inevitable that the NPT will be rocked by growing differences between the nuclear and non-nuclear weapon states.

The second section of the book is devoted to these troublesome issues of non-proliferation. Vertical proliferation amongst the nuclear armed states, as well as the cases of horizontal proliferation—of Iran and the possibilities in East Asia—are considered in detail in specific chapters. The continuing concern of nuclear terrorism and the state of the NPT are also examined to provide a holistic picture. A lacuna of this section is the non-inclusion of a chapter on North Korea, which unfortunately could not be obtained. However, the overall mood of pessimism on non-proliferation is palpable as the world moves down the third decade of the 21st century.

**Nuclear Disarmament**

Seventy-six years after the very first resolution of the United Nations General Assembly (UNGA) called for elimination of "national armaments on atomic weapons", the goal of disarmament still remains elusive. The focus of the NPT has always remained on non-proliferation. The NNWS have never been able to demand any serious actions from the NWS on their commitment on Article VI of the treaty. Some steps agreed to at a few NPT RevCons have proven to be merely tactical concessions for the strategic objective of keeping the treaty together.

It may be recalled that in the run-up to the NPT RevCon in 2010, there was some enthusiasm on nuclear disarmament. A set of op-eds between 2007-09 by the 'four horsemen', hardcore US Cold Warriors, George Schultz, Sam Nunn, Henry Kissinger, and William Perry warned of the danger of a nuclear apocalypse and argued in favour of a world without nuclear weapons. Barack Obama, in his capacity as the American president, added to this hope by presenting his own vision of such a world, even if he cautioned that it was distant.

The distance, in fact, lengthened as relations between the USA and Russia deteriorated around the middle of the 2010s. Meanwhile, the rise of a belligerent and assertive China, armed with modern military and nuclear capabilities, put that brief moment of disarmament optimism to rest. It has not revived since then. Rather, nuclear modernisation in all nuclear armed states has been the flavour of the last decade. Trust deficits have grown and the arms control architecture has been dismantled, treaty by treaty. The New START (Strategic Arms Reduction Treaty) remains the only bilateral arms reduction agreement in force at this juncture. The hope held out by a Joint Statement announced by the P5 in January 2022 to collectively commit themselves to not fighting a nuclear war did not last even two months. Since the start of the war by Russia against Ukraine in end-February 2022, Russia's relations with the West now face a schism that will not be easy or quick to bridge. In fact, the prevalence of difficult relations is not confined to the US-Russia alone. All the other nuclear dyads too – the US-China, the US-North Korea, China-India, and India-Pakistan too are severely stressed. The chances of establishing mechanisms for strategic stability and confidence-building, leave alone arms control, remain bleak.

Frustration at this depressing reality, and an inability to get the NWS to move on disarmament, made the NNWS conclude a Treaty on the Prohibition of Nuclear Weapons (TPNW) at the UNGA in 2017. It entered into force in January 2021 and has managed a membership of over 60 countries. The irony remains, however, that all of these are non-nuclear states. So, they seem to have banned for themselves a weapon that they anyway do not possess. The NWS, meanwhile, have ignored this initiative. The first meeting of the states parties of the TPNW is scheduled for June 2022. While calls will be made to universalise the treaty, there is little hope that the NWS will pay attention to this exercise. Much scepticism on the prospects of nuclear disarmament prevails.

However, this situation could change. It could be that the nuclear armed states emerge out of the Russia-Ukraine conflict duly shaken by the nuclear risks and feel the need to take steps that lead up to eventual elimination of nuclear weapons. It could also be that populations become more aware of the dangerous consequences of the breakdown of nuclear deterrence, and demand action. Or, that some visionary leaders rise to the fore and are able to coalesce a critical mass on reducing the nuclear risks as a way station to disarmament. Alternatively, the world may continue to live dangerously with the nuclear risks!

In the hope of generating an informed discussion on this subject, the last section of the book is devoted to examining the consequences of the use of nuclear weapons, the concepts of Confidence-Building Measures

(CBMs) and arms control, and their implementation through treaties, as well as a detailed overview of the disarmament initiatives over the recent years. The final chapter undertakes a tentative exploration of the prospects of disarmament in the wake of the Russia-Ukraine conflict. The hope that good sense will prevail, and the fear that it may not, seem to coexist at this moment.

## Bringing the Book Together

Containing 15 chapters over the three disparate but interconnected nuclear issues, the book has been made possible by the hard work of each of the authors. A conscious attempt has been made to ensure diversity of gender and age amongst the contributors. Their expertise and professionalism are evident in their writings. It was a coup of sorts to be able to bring them all together in one book. I remain indebted to them for their effort, and the promptness with which they responded to our call from the Centre for Air Power Studies (CAPS).

It must also be mentioned that nearly all the articles were written before the start of the Russia-Ukraine conflict. While the nuclear landscape will feel the impact of this episode, the general trend lines will probably remain the same. So, there is likely to be a cautious optimism on nuclear energy, perhaps exacerbated by the dilemma posed to Europe by its excessive energy dependence on Russia. Non-proliferation will continue to evoke despondent pessimism, aggravated by a fear of proliferation if more countries feel tempted to acquire nuclear weapons for national security. Nuclear disarmament will remain in a state of suspended animation, and could swing either way, depending on how the conflict ends.

As a country that has a vested interest in all three aspects, India must invest its energies in proactively shaping the landscape. Hopefully, this book will serve to ignite thoughts and ideas amongst the practitioners, scholars, academics as well as general readers.

I am grateful to the Director General, CAPS, Air Marshal Anil Chopra, for sowing the seed of this book in my mind. His energy and enthusiasm are indeed infectious. Kalpana Shukla, our publisher, facilitated the highest support and standards at KW Publishers. Sushanta Gayen responded to my every requirement on the manuscript. Rehana Mishra used her 'eagle eye' and her 'way with words' to ensure greater readability of all the chapters. I am sure I speak for all the authors in expressing gratitude for our respective library staff in furnishing umpteen requests – and always with a smile. My gratitude also to family and friends who contribute in myriad indefinable ways. It goes without saying that I remain responsible for all errors of omission and commission in the book.

**Manpreet Sethi**

# Contributors

Ms **Zoya Akhter** is a foreign policy analyst, who did her Masters in Geopolitics and International Relations from Manipal University. She worked as a Research Associate at the National Institute of Advanced Studies, Bengaluru, and the Centre for Air Power Studies (CAPS), New Delhi. At CAPS, she completed a project on Nuclear Power in Asia, which is currently under publication. She writes regularly on the nuclear power policies of various countries.

Air Marshal **Anil Chopra** PVSM, AVSM, VM, VSM (Retd), is a fighter pilot, test pilot, and a pioneer of the Mirage-2000 fleet, who has commanded a Mirage 2000 squadron and the Flight Test Centre, Aircraft and Systems Testing Establishment (ASTE) of the Indian Air Force (IAF). He was the team leader of the MiG 21 Bison upgrade project in Russia. He has commanded operational air bases in both the Western and Eastern Sectors. Air Marshal Chopra was head of the IAF in Jammu and Kashmir ( J&K), and head of Operational Inspections of the IAF. He retired as Air Officer Personnel. He has been a member of the Armed Forces Tribunal, Lucknow Bench, and member of the Executive Council of Jawaharlal Nehru University (JNU), and is an avid writer and author. Currently, he is the Director General of the Centre for Air Power Studies (CAPS).

Dr **Silky Kaur** is an Associate Fellow at the Centre for Air Power Studies (CAPS), New Delhi. She obtained her PhD in Nuclear Strategic Studies from the Diplomacy and Disarmament Division,

Jawaharlal Nehru University (JNU), New Delhi, in 2020. Her research interests include nuclear strategy, energy, proliferation, arms control and disarmament. At CAPS, she is working on a book project titled "Emerging Technologies and Nuclear Deterrence", which explores the impact of emerging and disruptive technologies such as new missiles and missile defences, artificial intelligence, unmanned aerial and underwater vehicles on nuclear deterrence.

Dr **Reshmi Kazi** is a Professor at Jamia Millia Islamia, New Delhi, where she teaches arms control and nuclear disarmament. She was an Associate Fellow at the Manohar Parrikar Institute for Defence Studies and Analyses, New Delhi (2007-17). She has written and presented extensively on nuclear security issues, including a paper on "Nuclear Terrorism and UN Resolution 1540: A South Asian Perspective" at the United Nations Headquarters, New York. Her latest publication includes the co-edited volume *India and Global Nuclear Governance*, published by Routledge.

Ms **Nasima Khatoon** is a Research Associate at the Rashtriya Raksha University (RRU), India. Earlier, she worked as a Research Fellow at the Centre for Air Power Studies (CAPS), New Delhi, and the National Institute of Advanced Studies (NIAS), Bengaluru. Her areas of research are India's foreign policy, nuclear strategy and security, and borderland studies of northeast India. She holds a Master's degree in Defence and Strategic Studies from the Savitribai Phule Pune University, India. She has written book chapters, journal papers and short analytical articles for Springer, Routledge, *Air Power Journal*, *The Hindu* and *Diplomat*, among others.

Lieutenant General **Prakash Menon** is a former Military Adviser, National Security Council Secretariat and Commandant, National Defence College. Presently, he is the Director, Strategic Studies Programme, Takshashila Institution, Bengaluru, Professor Emeritus, Transdisciplinary University, Bengaluru, and Adjunct Professor,

National Institute of Advanced Studies, Bengaluru. He is the author of *Strategy Trap: India and Pakistan Under the Nuclear Shadow*.

Dr **Sitakanta Mishra** is currently an Associate Professor of International Relations in the School of Liberal Studies (SLS) of Pandit Deendayal Energy University (PDEU), Gujarat, India. Earlier, he worked as a Research Fellow at the Centre for Air Power Studies (CAPS), New Delhi. He is also the Managing Editor of the *Liberal Studies* journal and former Associate Editor of the *Indian Foreign Affairs Journal* (Association of Indian Diplomats, New Delhi). Dr Mishra specialises in nuclear safety-security, missiles, national security and global affairs.

Dr **Jagannath Panda** is the head of the Stockholm Centre for South Asian and Indo-Pacific Affairs (SCSA-IPA) at the Institute for Security and Development Policy (ISDP), Sweden. He is also the Director for Europe-Asia Research Cooperation at the Yokosuka Council on Asia-Pacific Studies (YCAPS) in Japan; a Senior Fellow at The Hague Centre for Strategic Studies (HCSS), The Netherlands; an International Research Fellow at the Canon Institute for Global Studies (CIGS) in Japan; and a Senior Fellow at the United Services Institute of India. He is also the Series Editor for *Routledge Studies on Think Asia*. Dr. Panda is the author of the books *India-China Relations* (Routledge: 2017) and *China's Path to Power: Party, Military and the Politics of State Transition* (Pentagon Press: 2010). Dr. Panda's recent edited works are: *Chinese Politics and Foreign Policy under Xi Jinping* (co-edited), *Quad Plus and Indo-Pacific* (Routledge, 2021); *Scaling India-Japan Cooperation in Indo-Pacific and Beyond 2025* (KW Publishers Ltd., 2019), and *The Korean Peninsula and Indo-Pacific Power Politics: Status Security at Stake* (Routledge, 2020).

Dr **R. Rajaraman** is currently Emeritus Professor of Physics at Jawaharlal Nehru University (JNU). His PhD in Theoretical Physics (Cornell, 1963) was followed by research/faculty positions at Cornell

University, Institute for Advanced Study at Princeton, Harvard, Stanford, Massachusetts Institute of Technology (MIT), CERN (*Conseil Européen pour la Recherche Nucléaire*) (Geneva) and Indian Institute of Science (IISc), Bengaluru. He has also worked on global arms control, India's nuclear doctrine, and nuclear energy. His work on disarmament has been recognised internationally through the prestigious Leo Szilard Award and, in physics, in numerous ways, including India's SS Bhatnagar Prize. He is a former co-chair of the International Panel on Fissile Materials and former Vice-President of the Indian National Science Academy (INSA).

Dr **Kanica Rakhra** is a former consultant with the Disarmament and International Security Affairs Division of the Indian Ministry of External Affairs. She completed her PhD from the Centre for International Politics, Organisation and Disarmament, Jawaharlal Nehru University (JNU) in 2016, and has undertaken Research Fellowships such as the Visiting Fellowship at Middlebury Institute in Monterey (2019) and the Harvard-led Arms Control Academy (ACONA) Fellows (2020-21), a collaborative effort of six think-tanks and institutions from Europe, Russia and the United States of America. Currently, she is a consultant with the Research and Information System for Developing Countries (RIS), New Delhi.

Mr **Abhishek Saxena** is a Research Associate (Nuclear Studies Group) at the Centre for Air Power Studies (CAPS). His research interest focusses on nuclear strategy, nuclear proliferation, and the implications of emerging technologies on national security. Before joining CAPS, he worked as a Research Intern at the Observer Research Foundation (ORF), New Delhi. He has a Master's degree in Politics and International Studies from the School of International Studies, Jawaharlal Nehru University (JNU), New Delhi, and B.Sc. (Hons) in Chemistry from Kirori Mal College, University of Delhi.

Dr **Sheel Kant Sharma** is a Distinguished Alumnus of the Indian Institute of Technology (IIT), Bombay, which awarded him a PhD in Physics (1975). He joined the Indian Foreign Service in 1973, served in Kuwait, Saudi Arabia, Geneva, Algeria and Vienna, and headed important divisions in the Ministry of External Affairs like the Disarmament and International Security Affairs (DISA) and Southern Division as a joint secretary. He was Indian Ambassador to Austria, and Permanent Representative of India to the UN in Vienna and the International Atomic Energy Agency (IAEA). As Indian Governor, he served on the IAEA Board of Governors from 2004 to 2008. Dr Sharma was the last Indian Secretary General of the South Asian Association of Regional Cooperation (SAARC) from 2008 to 2011. He writes regularly on disarmament, nuclear policy, non-proliferation and deterrence, South Asian regionalism, and Association of Southeast Asia Nations (ASEAN), and is noted for commentaries on television and All India Radio.

Dr **Swaran Singh** is a Professor and Chair, Centre for International Politics, Organisation and Disarmament at the School of International Studies, Jawaharlal Nehru University (JNU), New Delhi. He is president of the Association of Asia Scholars, and a regular commentator on Asian affairs on audio, visual, social and print media. Among his most recent published works are a research article titled "Mekong-Ganga Cooperation: Interest, Initiatives and Influence", published in *Asia Policy*, Vol. 17, April 2022, and a co-edited book titled *Revisiting Gandhi: Legacies for World Peace and National Integration* (Singapore: World Scientific, 2022).

# Introduction

The global nuclear landscape is characterised by developments of nuclear weapons, non-proliferation, disarmament initiatives, and, very importantly, nuclear energy. The capabilities around existing nuclear weapons are improving, the delivery systems are becoming fast, precise and with longer ranges. Some countries have begun increasing the number of their weapons or warheads. Countries are also developing nuclear weapons with smaller yields, implying possible tactical or coercive impacts on the battlefield. There continue to be enduring security threats to weapons and material, and the risk of proliferation.

After the Ukraine conflict, more and more countries are likely to want nuclear weapons on their soil to act as a deterrent. New nuclear weapons are being developed without nuclear tests. Some countries like China and North Korea are increasing their stockpiles of nuclear weapons. China, while concentrating on mobile missiles with Multiple Independently Targeted Reentry Vehicles (MIRVs), is also building new nuclear missile silo parks. To counter sophisticated ballistic missile air defence systems like those with the USA, both China and Russia are developing hypersonic glide vehicles with possible nuclear warheads. Russia and China are looking at nuclear-armed, nuclear-powered underwater unmanned vehicles. Among all this, there is the threat of theft or diversion of fissile materials. There are nuclear safety issues also.

Both vertical and horizontal proliferation is on the rise. Great power nuclear competition and the arms race have begun again. The Russians and Chinese are modernising their arsenals to counter the USA's large conventional and nuclear counter-force capability and sophisticated ballistic missile defence system. The US Administration is already threatening to respond with modernisation of its nuclear arsenal. Withdrawal from the Intermediate Nuclear Forces (INF) Treaty, and questions on the future of the New Stategic Arms Reduction Treaty (START) after 2026 further threaten to disturb strategic stability. The conflict in Ukraine has caused serious wounds that will take time to heal. There is an increasing number of players involved in horizontal proliferation. Iran continues to pursue nuclear power, and, perhaps, weapons. South Korea and Japan want a more assured nuclear umbrella. Saudi Arabia wants to acquire nuclear weapons to counter the Iranian programme.

The Ukraine conflict has also seen the nuclear line being pushed. Nuclear powered China and India have had a bloody skirmish. India's Balakot strike, albeit with limited objectives, was against a nuclear armed state. The nuclear powers are increasingly running the risk that they may stumble into a war, and there are uncertainties. North Korea's Kim Jong Un's regular missile testing is keeping the world on tenterhooks. Led by an autocratic, very ambitious leader, China's nuclear weapons, delivery platforms, and doctrinal evaluation are being advanced and monitored closely. Though rising, China's strategic nuclear force is still way behind the United States and Russia. China has put many other capabilities in place before nuclear modernisation. China's massive conventional short-range ballistic missile capabilities comprise a big asset in tactical battle scenarios. Advances in Artificial Intelligence (AI) will support decision-making and the command and control matrix. No arms control treaty can move forward as long as there is a feeling of asymmetric advantage, and till China is brought on board.

While there is a massive drive to get away from fossil fuels toward green energy, in its 2020 edition of Energy, Electricity and Nuclear Power Estimates for the Period up to 2050, the International Atomic Energy Agency's (IAEA's) high case projection has global nuclear generating capacity increasing from 392 Gigawatts electric (GWe) in 2019 to 475 GWe by 2030, 622 GWe by 2040 and 715 GWe by 2050. The amount of energy provided by nuclear generation will be nearly double. To achieve this, new capacity additions would reach 30 GW per year in the early 2030s. However, the World Energy Council (WEC), under its main scenario, where 'intelligent' and 'sustainable' economic growth models emerge as the world seeks a low-carbon future, shows that nuclear accounts for 17 per cent of electricity generation in 2060, from a global installed capacity of 989 GWe. With over 22 nuclear reactors in seven power plants, India's current nuclear power capacity of 6,790 Megawatts (MW) is expected to increase to 22,480 MW by 2031. This, in turn, is going to assist the country in meeting low carbon targets, along with other clean energy sources. Future widespread use of electric vehicles, both pure electric and plug-in hybrids, will increase electricity demand, especially at night.

Small Modular Reactors (SMRs) have begun seeing action. The Floating Nuclear Power Plant (FNPP) operating in Pevek in Russia's Far East, was the first to become operational in May 2020. The construction of the world's first commercial land-based SMR started in July 2021 with the Chinese power plant Linglong One. The operation of this prototype is due to start by the end of 2026. SMRs save on staffing, security and deployment time. Yet the licence processing is slow because of nuclear safety and security concerns. SMRs have great potential for use in out-of-area-contingencies and in remote places. Governments and industry need to accelerate the development of SMRs. The military uses of SMRs are in nuclear powered ships and submarines. Mobile nuclear plants are carried to operational areas. Soon there will be applications of nuclear power in space ships and batteries.

The Treaty on Non-Proliferation of Nuclear Weapons (NPT) is regarded as the cornerstone of the global nuclear non-proliferation regime and an essential foundation for the pursuit of nuclear disarmament. The NPT non-nuclear weapon states agree never to acquire nuclear weapons and the NPT nuclear weapon states in exchange agree to share the benefits of peaceful nuclear technology and to pursue nuclear disarmament aimed at the ultimate elimination of their nuclear arsenals. With the five authorised nuclear weapons states still having 13,400 warheads in their combined stockpile, critics argue that the NPT cannot stop the proliferation of nuclear weapons or the motivation to acquire them. It is hoped that the five would make "good faith" efforts to reduce and ultimately eliminate their nuclear arsenals. Current arsenals are big, but they are only about one-fifth of what they were a half-century ago. The arsenals of the United States and Russia still account for more than 90 per cent of the total number of warheads on Earth today. The United Kingdom, France, China, India, and Pakistan each likely possesses between 150 and 300, with Israel's unconfirmed arsenal totalling almost 100 bombs. North Korea probably has a couple dozen nuclear weapons, with enough fissile material to make a few dozen more. There will be need to guard against nuclear accidents, manage nuclear waste, and prevent proliferation.

Deterrence theory holds that nuclear weapons are intended to deter other states from attacking with their nuclear weapons, through the promise of retaliation and possibly mutually assured destruction. For nuclear deterrence to function properly, three important aspects are necessary: one, that a state's nuclear arsenal must appear to be able to survive an attack by the adversary and be used for a retaliatory second strike; two, the state must not respond to false alarms of a strike by the adversary; three, the state must maintain command and control.

An era of new great power competitions has already begun. The United States remains a superpower, still ahead of the rest, followed by China, which is an "emerging superpower," and Russia remains a great power, primarily because of its nuclear arsenal. In February 2018, the US released the Nuclear Posture Review (NPR), which seeks to ensure that the US will continue to maintain a safe, secure, and effective deterrent that protects it and its allies, and above all, deters adversaries. It addresses the importance of nuclear deterrence; pragmatic arms control and non-proliferation initiatives; enhancing strategic stability with potential adversaries such as Russia and China; and putting in place effective oversight mechanisms to address cost concerns associated with the strategic nuclear modernisation programme.

The Iran nuclear deal was running reasonably smoothly with adequate checks and balances, till, in 2018, the Trump Administration withdrew from the deal, reapplied the sanctions, and removed the oil export waivers. Seeing it as a breach of the deal, Tehran started exceeding the agreed-upon limits to its stockpile of low-enriched uranium, and began enriching uranium to higher concentrations, albeit still far short of the purity required for weapons. Negotiations took a little positive turn with the Biden Administration, but the talks halted again after Russia's invasion of Ukraine complicated the discussions. Tehran and Washington still disagree on several issues. The oil and gas embargo on Russia once again brings hope for a deal with Iran.

Australia, the United Kingdom, and the United States trilateral security partnership (AUKUS) is meant to deliver nuclear-powered submarines to the non-nuclear Australia. Australia will become the first non-nuclear weapon state to exercise a loophole that allows it to receive nuclear propelled submarines under the NPT and the inspection system of the International Atomic Energy Agency (IAEA). Many are saying that the AUKUS submarine deal is bad for nuclear non-proliferation and will set a new precedent.

As China modernises and expands its nuclear arsenal in competition with the United States, India's concerns begin to rise and that has a cascading impact on Pakistan. The US-India civil nuclear deal and establishment of a strategic partnership resulted in China offering a similar arrangement to Pakistan. The complex and fluid nuclear environment is not good. It is perhaps time for a genuine dialogue, risk reduction, and restraint.

Meanwhile, the threat of nuclear materials or weapons being acquired by terrorists remains high. The 2005 United Nations International Convention for the Suppression of Acts of Nuclear Terrorism is meant to address such risks. Al-Qaeda, the Islamic State, North Caucasus terrorists, and the Japanese terror cult Aum Shinrikyo have made failed attempts to acquire a "dirty bomb". The traditional deterrence is not an effective approach toward terrorist groups bent on causing a nuclear catastrophe. Finding means of tracing nuclear material, and preventing theft is important. Countries like Pakistan are high risk sources for such thefts.

This book on "The Global Nuclear Landscape—Energy, Non-proliferation, and Disarmament" looks at a wide variety of subjects and dynamics. It covers developments in nuclear energy and also its growth in India. Nuclear non-proliferation has been dealt with in great detail. Finally, the book looks at nuclear deterrence and disarmament. It contextualises all the issues through India's aspirations and challenges and tries to set the direction on India's compass. My compliments to Dr Manpreet Sethi for putting together a wide canvas of subjects, written by well-known analysts of nuclear issues. The book is a compendium on all nuclear issues. It makes a great read and is a must-have for all security practitioners.

Air Marshal **Anil Chopra** PVSM AVSM VM VSM (Retd)
Director General
Centre for Air Power Studies

# NUCLEAR ENERGY

# 1. Current Nuclear Energy Reality: Drivers and Challenges

*Manpreet Sethi*

Eleven years after the nuclear accident at the Japanese nuclear power station at Fukushima impacted public confidence and government policies on nuclear energy, it is time to reassess the current nuclear energy reality. How much have the figures on nuclear energy generation changed from 2012 to 2022? In which countries? Why have some countries like China and India continued to stay the course on their nuclear energy targets? What are the contemporary drivers for nuclear power programmes? And, what are the major challenges that still stand in the way? This paper addresses these questions to provide a realistic assessment of the global picture on nuclear energy.

## Nuclear Energy over the last Decade: A Global Snapshot

According to the International Atomic Energy Agency (IAEA) Power Reactor Information System (PRIS)[1], in 2011, the year the accident at Fukushima happened, the total production of nuclear energy was 348 Gigawatts electric (GWe). This accounted for 13.8 per cent of the total global share of electricity generation. Some 448 nuclear reactors were then operable across the world and 61 more were under construction. In response to the accident, Japan immediately shut down 54 of its operating nuclear reactors. Many other countries too suspended operations at their nuclear plants as they conducted safety

1. "Nuclear Power Capacity Trend", *IAEA World Statistics*. https://pris.iaea.org/PRIS/WorldStatistics/WorldTrendNuclearPowerCapacity.aspx. Accessed on February 14, 2022.

audits of their own. Some even announced a phase-out of their nuclear power programmes, while some that were considering starting such programmes, decided to pause.

As a result, in 2012, the global electricity generation had dropped by 7 per cent. The World Nuclear Performance Report, published by the World Nuclear Industry Status Report (WNISR), in 2013 said, "The nuclear industry is in decline: the 427 operating reactors are 17 lower than the peak in 2002… About three-quarters of this decline is due to the situation in Japan, but 16 other countries, including the top five nuclear generators, decreased their nuclear generation too."[2] Nuclear sceptics were ready to write off nuclear energy.

However, the situation began to slowly improve by the middle of the 2010s. The period between 2013 and 2020 marked a year-on-year increase in nuclear generation. The year 2020 ended with global nuclear output of 392.6 GWe from 442 operational nuclear power reactors in 32 countries; and 27 more states were actively considering, planning or embarking on a nuclear power programme.[3] Overall, nuclear power capacity has gradually increased with the addition of 23.7 GWe from either the new units connected to the grid or upgrades to existing reactors.[4] In 2021, nuclear electricity production remained at a little over 390 GWe[5] from 439 operational reactors, with 52 more under construction.[6]

In 13 countries in 2020, nuclear electricity contributed 25 per cent to their total national electricity share. In France, three quarters of the electricity came from nuclear energy. Slovakia and Ukraine got

2. WNISR, "World Nuclear Industry Status Report: Executive Summary and Conclusions", July 11, 2013.
3. IAEA, *Annual Report 2020* (Vienna: IAEA, 2020), p. 41.
4. Marta M. Gospodarczyk, "Nuclear Power Proves its Vital Role as an Adaptable, Reliable Supplier of Electricity during COVID-19", *IAEA News Centre*. Available at https://www.iaea.org/newscenter/news/nuclear-power-proves-its-vital-role-as-an-adaptable-reliable-supplier-of-electricity-during-covid-19.
5. n. 1.
6. "Under Construction Reactors", *IAEA World Statistics*. https://pris.iaea.org/PRIS/WorldStatistics/UnderConstructionReactorsByCountry.aspx. Accessed on February 15, 2022.

50 per cent of their electricity from nuclear plants.[7] But it is in Asia that new nuclear capacity build is most active. In 2019, over 77 per cent of all new capacity additions, and in 2020, 36 per cent of new capacity additions have been in Asia. 2 GWe was connected to the grid in China in 2020. In the same year, Belarus and the United Arab Emirates (UAE) made their first nuclear reactors operational.

As of 2022, the largest amount of new nuclear build is taking place in China, with 18 reactors being simultaneously constructed. In the last decade, China brought online more than 36 GWe of new nuclear capacity.[8] Most of the new units were built in around five years, except for the first European Pressurised Reactor (EPR) (being built by the French) and AP1000 (being built by the USA). China is currently producing about 50 GWe from 50 operational nuclear reactors, which makes up for about 5 per cent share of the total national electricity production.

After China, India comes a close second on nuclear construction, with 7 reactors at different stages of completion. The present installed nuclear power capacity in the country is 6,780 Megawatts electric (MWe) and the share of nuclear power in the total electricity generation was about 3.1 percent in the year 2020-21. The government has approved financial sanction for 10 indigenous Pressurised Heavy Water Reactors (PHWRs) with an aggregate capacity of 7,000 MW. These will be built in fleet mode over this decade, with an expectation that one will be connected to the grid every year over the next ten years. The target is to increase nuclear power capacity to 22,480 MW by 2031.[9]

---

7. World Nuclear Association, *Nuclear Power in the World Today*, February 2022. Available at https://world-nuclear.org/information-library/current-and-future-generation/nuclear-power-in-the-world-today.aspx#:~:text=Nuclear%20energy%20now%20provides%20about,in%20about%20220%20research%20reactors. Accessed on February 15, 2022.
8. World Nuclear Association, *Nuclear Power in China*, updated January 2022. Available at https://world-nuclear.org/information-library/country-profiles/countries-a-f/china-nuclear-power.aspx.
9. Surendra Singh, "Govt Targets to Raise India's N-power Capacity to 22,480 MW from 6,780 MW by 2031, *The Times of India*, May 24, 2022.

Russia too maintains a high tempo in its nuclear sector and is constructing a new generation of standardised large reactors to replace older units. In 2021, it also became the first country to deploy a floating nuclear power plant aboard the *Akademik Lomonosov*. Russia today holds a large chunk of the global nuclear exports. Over 20 nuclear reactors are confirmed for construction abroad.[10] On the domestic front, it is producing almost 28 GWe from 37 nuclear reactors, with three more under construction.

Meanwhile, South Korea too has its own line of standardised large reactors: 24 reactors provide about a third of the electricity in the country. Political compulsions made it turn away from domestic expansion of its nuclear programme in 2018, but the arrival of a new government after elections in April 2022 is likely to bring in a return to nuclear energy given the professed inclinations of the new leaders.

In France too, President Macron has indicated a desire to move back to full scale nuclear development. On February 10, 2022, he outlined a new energy policy for carbon free electricity production which includes construction of six new nuclear reactors, consideration of building a further eight and a push ahead with the development of small modular reactors.[11] Emphasising the need for renewables and nuclear as carbon free sources of electricity, Macron has described nuclear as "the most relevant choice from an ecological point of view and the most expedient from an economic point of view and, finally, the least costly from a financial point of view."[12]

The USA brought a new reactor into operation in 2016 after a long hiatus of nearly two decades. During this period, more reactors were brought to early closure. Availability of cheap natural gas, subsidies for renewable sources, and politics were among the reasons for the apparent deprioritisation of nuclear energy. Despite that, though, given the large

10. World Nuclear Association, *Nuclear Power in Russia*, updated December 2021. Available at https://world-nuclear.org/information-library/country-profiles/countries-o-s/russia-nuclear-power.aspx.
11. "Macron Sets out Plan for French Nuclear Renaissance", *Modern Power Systems*, February 15, 2022.
12. Ibid.

number of nuclear reactors built in the USA between the 1950s and 1980s, the country still has the largest number of operational nuclear reactors and produces the largest amount of nuclear energy at nearly 790 Terawatt hours (TWh). More recently, the emphasis on nuclear energy seems to be returning as part of the transition to clean energy. According to the US Office for Nuclear Energy, "The US Department of Energy (DOE) spent a record $1.3 billion to support nuclear energy research in 2021 and is receiving billions more through the Bipartisan Infrastructure Law to support the current US reactor fleet and help deploy new advanced reactor designs within the decade."[13] As part of this legislation, $6 billion has been budgeted for a new civil nuclear credit programme that will help nuclear operators apply for certification and offer bids for credits to support continued operations, and avoid any premature retirement owing to financial reasons.

The above has been a brief overview of some of the select major nuclear energy producers across the world. The current global share of nuclear electricity is at 10 per cent owing to an increase in overall electricity production, mostly from the addition of new solar and wind capacities. Fig 1 provides a glimpse of the share of different fuels in world electricity production.

**Fig. 1: World Electricity Production by Source 2021**

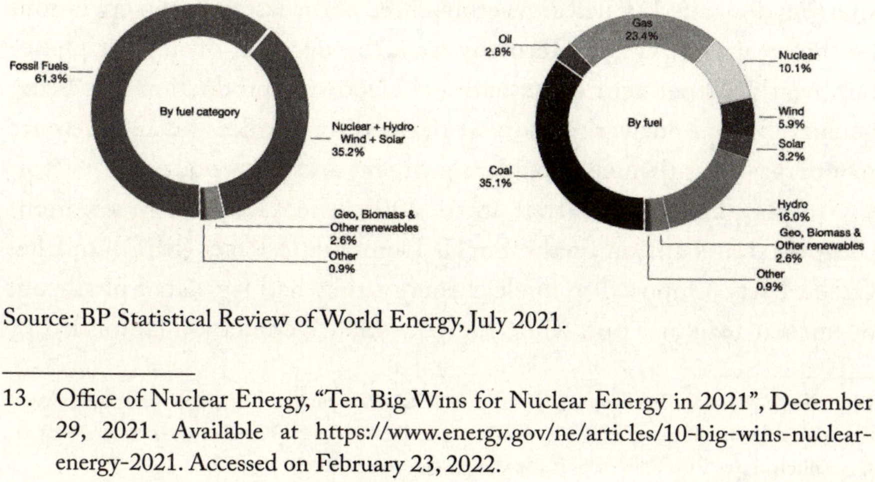


Source: BP Statistical Review of World Energy, July 2021.

13. Office of Nuclear Energy, "Ten Big Wins for Nuclear Energy in 2021", December 29, 2021. Available at https://www.energy.gov/ne/articles/10-big-wins-nuclear-energy-2021. Accessed on February 23, 2022.

As is evident, nuclear power has steadily chugged along despite Fukushima. According to a report on global nuclear energy trends, "more than half of the reactors permanently shut down in the last few years have done so not because of technical limitations, but because of political phase-out policies or the failure of markets to adequately recognize the value of on-demand, low-carbon, reliable nuclear power."[14] The following section explains some of this politics that bedevils national energy policies.

## Understanding National Electricity Choices

Countries choose the sources from where to generate electricity based on their unique circumstances. Many factors determine these: national population statistics, level of economic development, projected electricity requirements, availability or non-availability of fuel sources, implications of geopolitical conditions on overseas fuel imports, and sensitivity of public opinion towards the impact of electricity generation on human health and the environment. Owing to the distinctiveness of each country's position on these issues, their energy mix varies. This fact is best demonstrated in the disparate choices made by Germany, China and India in response to the accident at Fukushima.

In 2011, Germany decided to phase out nuclear energy by 2022. Moving towards this objective over the last decade, the country has been shutting down its 17 nuclear reactors: three of the last will stop operations by this year end. When Germany took the decision of nuclear phase-out, nearly 22 per cent of its national electricity production was being generated from nuclear reactors. It decided to sacrifice this in deference to a pervasive anti-nuclear public sentiment and party politics.

It may be recalled that until 2007, the German government comprised a coalition of the Social Democratic Party (SPD) and the Green Party. Opposed to nuclear energy, they had legislated phase-out of nuclear reactors. But, when the Christian Democratic Union led by

14. *World Nuclear Performance Report 2021: COP26 edition*. Available at https://www.world-nuclear.org/our-association/publications/global-trends-reports/world-nuclear-performance-report.aspx. Accessed on January 15, 2022.

Angela Merkel came to power in 2007, she reversed this decision and let reactors continue as long as they could meet the safety criteria. Post Fukushima, however, she too was compelled by the public mood after the loss of her party in a pivotal parliamentary election. This defeat was interpreted as public rejection of nuclear energy. Accordingly, the German Cabinet approved the closing down of all plants in a decade.

Besides the political dimension of the decision, the socioeconomic realities of Germany too played a role in pushing the government to abandon nuclear power. Factors such as a stable population, a mature electricity market, high per capita energy availability of 7,030 Kilowatt hours (KWh), projections of marginal increase of electricity, and Germany's position as an exporter of electricity, helped the country to take this step. It also was fortuitous that Germany was technologically advanced enough to step up its efforts towards rapid exploitation of renewables to make this transition. And in the interim, it knew it had the cushion to make up any shortfall through imports of coal from Poland, gas from Russia, or even electricity from France or Czechoslovakia.

Over the last decade, Germany has seen significant deployment of renewables, their share having grown from 25 per cent in 2013 to 45 per cent in 2021.[15] However, the country's dependence on Russia for gas has also grown. Meanwhile, the costs of the nuclear phase-out have not been trivial. On the financial front, the country has had to pay an amount of US$3 billion to utilities that were forced into premature closure, and the cost of electricity for consumers in Germany has increased.[16] On the environmental front, increased fossil fuel consumption has led to raised pollution levels. But, most importantly, a geopolitical cost is being paid through increased dependence on Russia, which has compelled Germany to view the Russian threat being posed to Ukraine differently from the

15. "Renewable Energies in Figures", November 2021. Available at https://www.umweltbundesamt.de/en/topics/climate-energy/renewable-energies/renewable-energies-in-figures. Accessed on February 10, 2022
16. R B Grover, "An Examination of the Narratives about the Electricity Sector", *Current Science*, Vol 119, No. 12, December 25, 2020

position of its other North Atlantic Treaty Organisation (NATO) and European Union (EU) allies.

In contrast to Germany, China and India were producing only 4 per cent and 3 per cent respectively of their electricity from nuclear plants in 2011. In response to Fukushima, both countries conducted safety assessments, tempered their nuclear targets but chose to renew their nuclear commitments. Their decision to stay with nuclear energy was determined by the fact that both countries have growing populations, high electricity demand projections[17] owing also to the rapid economic growth and infrastructure development, fairly low per capita availability of electricity at just about 1,300 KWh in the case of India, and 3,400 KWh in the case of China, and limited availability of other fuel sources. Not surprisingly then, the two Asian powers have found it prudent to expand their nuclear programmes rather than abandon them despite Fukushima. Country choices on electricity sources, thus, can be very differently affected by the same trigger event owing to exclusive situations.

### Contemporary Drivers of Nuclear Energy

In the 1950s, rapid industrialisation and economic growth across major economies led to the adoption of sources of energy that were cheap, reliable and quickly deployable. Therefore, coal and other fossil fuels became the obvious choice for electricity generation. This situation changed in the 1970s when the oil shock brought the need for energy security into sharper focus. A desire to reduce vulnerability to energy imports then became the major driver for nuclear power programmes. Japan and France were among the first to move in this direction for this reason.

The more contemporary driver for nuclear energy, however, is the need for a low carbon energy source. Given the continuing high demand

17. India is the third largest energy consumer in the world and has the highest share of global energy demand over the next few decades. "India at the Centre of the World Energy Stage: IEA Director", *Down to Earth*, January 2021. Available at https://www.downtoearth.org.in/news/energy/india-at-the-center-of-world-energy-stage-will-experience-gains-iea-director-75499. Accessed on February 23, 2021.

for electricity (nearly a billion people globally still do not have access to electricity), and the growing climate change concerns, the challenge now being faced is to produce more electricity from sources that cause the least impact on the environment. The Energy Impact Assessment of 2019 has projected a 50 per cent increase in world energy usage by 2050, most of it in countries of Asia. As explained by Bill Gates, "The global population is headed toward 10 billion by end of the century, and much of this growth is happening in cities that are highly carbon intensive".[18] As more people live longer and healthier lives, with aspirations for a higher standard of living, there is also a rising demand for infrastructure and household goods. Electricity is needed to power them all.

However, across nations there is a greater sensitivity to the impact of human activities on health and the environment. Interestingly, every activity of man, including breathing, contributes carbon dioxide ($CO_2$) to the atmosphere. However, some activities cause much higher greenhouse gas (GHG) emissions than others. For example, making cement and steel or plastic contributes 31 per cent of carbon emissions into the atmosphere. Electricity generation comes a close second at 27 per cent. Agriculture and animals contribute 19 per cent, and transportation 16 per cent to the total GHG emissions.[19]

Not only is electricity generation amongst one of the highest contributors to air pollution, it is also one factor whose transformation could have a bearing on all other activities. Consequently, greater focus on how electricity is produced could contribute to a significant extent towards addressing the increasing concerns about climate change. The link between GHG emissions and global warming is well recognised today as adverse natural events have increased in frequency and intensity. In order to address this growing risk, the Paris Agreement of 2015 had mandated that countries must take necessary national actions to limit global warming to below 2 degrees, preferably at 1.5 degrees, compared to pre-industrial levels. At the Glasgow Summit in November 2021,

18. Bill Gates, *How to Avoid a Climate Disaster: The Solutions We Have and the Breakthroughs We Need* (UK: Penguin Random House, 2021), p. 40.
19. Ibid.

countries articulated their nationally determined contributions to help decarbonise the economy. More recently, in February this year, the Intergovernmental Panel on Climate Change (IPCC) released its report on climate change that draws urgent attention to how human-induced climate change is posing a mounting threat to humanity. It states the grim fact that "even temporarily exceeding this warming level will result in additional severe impacts, some of which will be irreversible."[20]

As mentioned earlier, electricity generation is one of the major human activities that add to GHG emissions. In order to address this fact, several low carbon energy options are being adopted. Achieving energy efficiency is one of them. But this can be possible only in economies which already have high per capita electricity availability. Clean coal technology provides another option but it is expensive to retrofit in existing plants and also expensive to build such new plants. In any case, continued use of fossil fuels can still not get mankind to zero emissions.

Hydro energy offers another low carbon source, but it faces the problem of large-scale displacement and rehabilitation of people from areas that are submerged. In fact, another little known challenge that this poses is generation of a large amount of methane, another GHG, when large parts of carbon rich land are submerged under water for creating reservoirs.[21] Renewables like wind and solar have seen some heartening growth in the last decade, but they continue to suffer from the problems of intermittency, consequent need for costly back-ups; and the challenge of waste disposal of photovoltaic panels.

Nuclear energy in this context has some major advantages to offer. There is no doubt that nuclear electricity generation, in general, has helped reduce GHG emissions. The Organisation for Economic Co-operation and Development's (OECD's) energy agency has estimated

20. IPCC, *Climate Change 2022: Impacts, Adaptation and Vulnerability* (Geneva: WMO & UNEP, 2022). Summary for policy-makers, available at https://report.ipcc.ch/ar6wg2/pdf/IPCC_AR6_WGII_SummaryForPolicymakers.pdf. Accessed on March 1, 2022.
21. Ibid., p. 69.

that nuclear power generation averted the release of over 60 gigatons of $CO_2$ emissions that would have used coal or oil since 1980.[22] Fig 2, below, demonstrates the 'cleanliness' of nuclear energy as compared to all other sources of electricity generation.

**Fig. 2: Comparative Analysis of Energy Sources on their Safety and Environmental Friendliness**

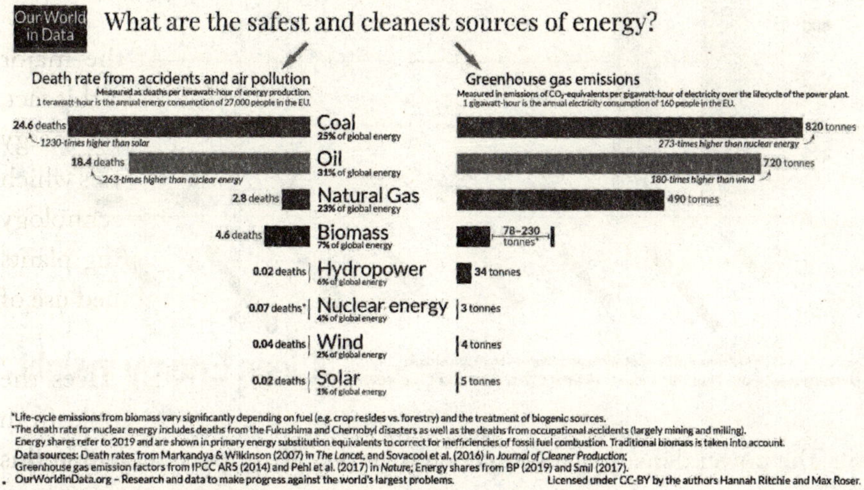


Source: https://ourworldindata.org/nuclear-energy#putting-death-rates-from-different-energy-sources-in-perspective

Besides, the fuel requirements of a nuclear power plant are much lower compared to a thermal plant. For instance, a 1,000 MWe coal-fired plant requires an annual amount of 38,000 railroad cars of coal, or 40,000 barrels of oil per day, while a nuclear plant of the same size only needs 6 truckloads of fuel. This also offers significant benefits by cutting down pollution caused by fuel transportation. The graphic in Fig 3 below provides an illustration of carbon dioxide emissions over the life-cycle of a source of electricity.

22. Chor Seng Tan, "Small Modular Nuclear Reactors", Standford.edu, December 11, 2015. http://large.stanford.edu/courses/2015/ph240/tan2/. Accessed on September 3, 2021.

Fig. 3: Comparison of Life-Cycle $CO_2$ Emissions

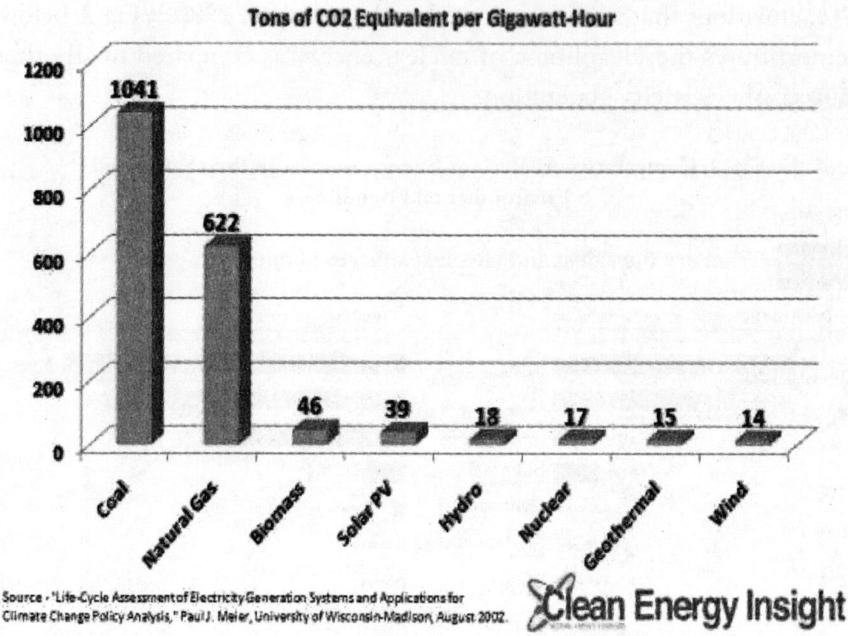


The power density of a source is also extremely important and on this front too, nuclear plants score above the other renewables. According to one estimate, "A solar farm needs between 5 and 50 times more land to generate as much electricity as an equivalent coal-powered plant, and a wind farm needs ten times more than solar."[23] Put another way, nuclear plants can generate 500-1,000 watts per sq m while solar is at 5-20 per sq m and wind at 1-2 watts per sq m.[24]

Gates emphasises that nuclear plants rank the highest in efficiently using materials like cement, steel and glass, which are electricity guzzlers themselves in their manufacturing processes. Nuclear plants use much less material per unit of electricity generated compared to the others.[25] In fact, solar photovoltaics are the least efficient, followed by hydropower,

23. Gates, n. 18, p. 96.
24. Ibid., p. 58.
25. Ibid., p. 85.

and wind. This is an important consideration when accounting for the GHG emissions produced in the manufacture of this material.

Another advantage of Nuclear Power Plants (NPPs) came to light during the COVID pandemic in 2020. Restrictions on economic and social activity during the lockdowns led to an extraordinary decline in the demand for electricity. In some countries, the demand reduced by as much as 10 per cent or more, compared to 2019 levels. This created challenges for both electricity generators and system operators. In these circumstances, nuclear power generation proved to be highly resilient, reliable and adaptable. This flexibility demonstrated by nuclear plants has triggered thinking on how it can be used to support a clean energy transition when used along with renewables. As said by an energy analyst, "Flexible NPP operation—or, in some cases, complete short-term shutdowns—supported grid operator needs and demonstrated nuclear power's ability to integrate into sustainable energy systems of the future, which are expected to rely significantly on flexible baseload generation technologies."[26]

As nations move towards increased use of renewables, the grids are likely to face greater instability, poorer power quality and increased incidence of blackouts. Nuclear power plants as a baseload source of electricity can fill this role. They can offer the advantage of minimising supply disruptions and sustainably support decarbonised electricity systems.

Independent studies undertaken by the Intergovernmental Panel on Climate Change (IPCC), World Nuclear Association (WNA), International Energy Agency (IEA) and Massachusetts Institute of Technology (MIT) have reached the conclusion that nuclear energy is a solution to the dilemma of how to increase electricity in an environmentally sustainable way. It has been stated that to achieve the target of limiting global warming to 1.5 degrees Centigrade, the share of nuclear energy would have to go up by 2.5 times of where it is today. This means that the share of nuclear electricity should be at 25 per cent

26. Gospodarczyk, n. 4.

as against the 10 per cent of today. However, this faces several challenges given some perceptions about nuclear energy.

## Challenges before Nuclear Energy

Public perception of the safety of nuclear power generation is a major challenge across countries. Despite the extremely few accidents that have occurred in nuclear plants, only three in seven decades of operations, the public view of nuclear energy remains dominated by fear and scepticism. Fig. 4 below maps the death rates across various sources of energy and it is evident that nuclear energy is close to the bottom of the table.

**Fig. 4: Death Rates from Energy Production**

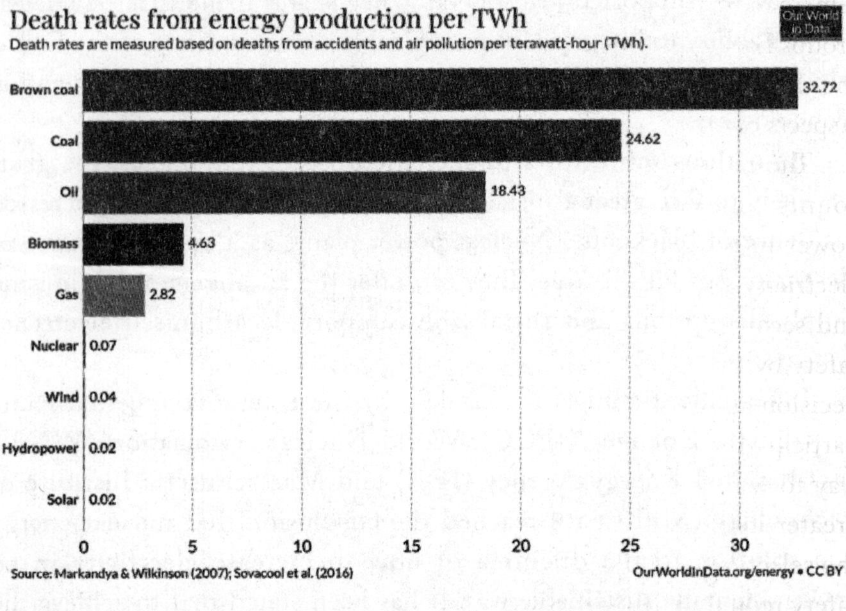


These statistics, however, are not well known. As Gates laments in his book, "Since the accidents at Three Mile Island and Chernobyl, America has broken ground on just two nuclear plants, even though more people die from coal pollution in a single year than have died in all nuclear accidents combined."[27] The sentiment runs across several nations and has not allowed domestic programmes to expand as much as could have been possible.

Two other factors too add to such a view. One of these is the high capital cost and long gestation period of nuclear reactor construction. Given that a majority of nuclear projects have faced cost and time overruns, the widespread perception of nuclear energy is that it is too expensive. The advantages of the long lifetimes of the plants, going up to 50 years in many cases, and the low operating cost of the reactors, owing to low fuel requirements are often overlooked.

The second issue that clouds public perception is the pending problem of how to ensure safe and secure spent fuel management. Nuclear waste management remains a key public concern since nuclear waste tends to remain radioactive for a very long time. Environmental and public interest groups have built their arguments against nuclear power by highlighting the difficulties involved in dealing with spent fuel and its safe storage. Aspects of safety and security are both involved in this dimension.

The battle of restoring public confidence has to be won on two counts: one, by making people understand the need for nuclear power in the global energy mix in order to ensure that generation of electricity does not adversely impact human and environmental health; and second, by explaining the high amount of importance accorded to safety by the nuclear industry. Unfortunately, in most countries, nuclear decision-making is an elite activity that pays little attention to public participation. But in the changed environment after Fukushima, the only way to win public support for nuclear energy will have to include a far greater interaction with the people to explain to them the reasons for the selection of a particular site, the basics of the reactor technology, the safety redundancies built into operations, etc.

In fact, in order to win over the public, nuclear plants should not hesitate to invite school and college students, organised groups of women's associations, the corporate sector, the media, and generally the common

27. Ibid., p. 47.

man, to visit the plants and raise their awareness on their many uses. A special effort must also be made to engage with Non-Governmental Organisations (NGOs) and local community groups, including religious heads, at plant sites since they have the advantage of directly interacting with the local populace as also a huge capacity for mobilising public opinion. The more approachable the nuclear plants seem, the greater will be the confidence that will be engendered over time.

Meanwhile, computing the cost competitiveness of nuclear energy cannot be a case of straight mathematics.[28] In fact, besides other issues, nuclear reactor construction has also been seen to become easy victim of political or public controversy. The issue of *political risk*, therefore, becomes a factor of consideration in the *economics* of the energy source. This then tends to increase the licensing or construction time, thereby adding to the scepticism that such plants can never rapidly add watts to the grid. This, in fact, remains a major inhibiting factor in endorsing nuclear energy to meet the climate change concerns. The fear remains that the plants may not be constructed in a timeframe wherein they can help mitigate the environmental challenges being faced right now.

However, in order to change public perception on this front, the nuclear industry must adopt the best construction management practices to reduce construction time as much as possible. In this context, it is noteworthy that there is an emerging trend towards Small Modular Reactors (SMRs) that, it is believed, would be able to address these challenges of traditional reactors. Chapter 3 of this book dwells upon SMRs in some detail. It is expected that reliable supply chains for factory-manufactured modular reactors will be able to cut down the time of construction and thereby keep the costs within projected estimates.

The governments can help in this regard by accepting nuclear energy as a solution to the contemporary problem of increasing electricity generation, but to do so with low carbon sources. They can also create an enabling environment by providing a level playing field on policy and finance, especially vis-à-vis renewable energy. This would also bring a

---

28. For a detailed discussion on this, see Grover, n. 16.

measure of predictability for the nuclear sector and incline the industry towards making investments that can provide for reliable supply chains. Therefore, it would be a virtuous cycle that could work to improve the economics of nuclear power.

On the aspect of nuclear waste, there is need for greater publicity for the fact that the nuclear industry is well aware of this challenge and has been managing waste disposal successfully for more than half a century. Dozens of facilities for low-level and intermediate-level nuclear waste are safely and securely in operation throughout the world. As far as the long-term management of high-level radioactive waste and spent fuel is concerned, nations are experimenting with the idea of construction of deep geological repositories. While the search for such sites is on in India as well, the country has also introduced the technology for reduction of high-level waste through separation of actinides and disposal of each, dependent on their chemical nature and composition. Given technological advancement and human ingenuity, it is quite possible that adequate solutions to the challenge of radioactive waste disposal will be found sooner rather than later.

The challenges facing nuclear energy are not insurmountable and it does offer a viable solution to the complex environmental issues confronting mankind. If an increase in electricity production is to be sustained with the least adverse impact to the environment, then nuclear energy has a role to play. In fact, enhancing energy production and sustaining the environment is not an either/or choice. The world needs both. Governments owe it to their citizens to provide them with electricity while also ensuring the best quality of life from their health and environmental perspectives. If economic development and decarbonisation have to be simultaneously pursued, nuclear power can offer a useful bridge that must be prudently built and safely operated.

# 2. India's Nuclear Energy Scenario

*Zoya Akhter*

## Introduction

India's civil nuclear programme is marked with distinct contradictions. On the one hand, the nuclear power sector receives reasonably strong political support across party lines, yet the expansion of nuclear power in the country has been slow, with nuclear energy contributing to just 3.1 per cent of the electricity generated.[1] While the importance of nuclear power in ensuring energy security and bringing down $CO_2$ emission levels in the country is reasonably well recognised, the slow growth can be attributed to several challenges that have been bedevilling the progress of the nuclear power sector such as long delays in project materialisation, shortages in financing, and public opposition, among others. Meanwhile, India has also made significant progress in the field of nuclear research and development and is currently developing an alternative nuclear fuel system, which, on completion, will not only reduce nuclear waste but is also expected to be more proliferation resistant. To understand these contradictions, this paper analyses the status of India's civil nuclear programme while assessing the various challenges hampering the development of the nuclear power programme and analysing the changing scenario in the Indian nuclear sector.

1. "India's Nuclear Power Capacity is Expected to Reach 22,480 MW by 2031 from current 6,780 MW: Jitendra Singh", Newsonair.com, August 4, 2021. https://newsonair.com/2021/08/04/indias-nuclear-power-capacity-is-expected-toreach-22480-mw-by-2031-from-current-6780-mw-jitendra-singh/. Accessed on September 28, 2021.

## Brief History and Overview of India's Nuclear Power Programme

India began to develop its civil nuclear programme in the 1940s as the vision of the post-independence leadership in the country focussed on attaining energy security to promote economic growth. Prime Minister Nehru in this regard reiterated the role of science and technology in nation building, what he famously referred to as the "temples of modern India." At the time when development policies were being formulated, Homi Bhabha, a nuclear physicist, who later became the founding director of the atomic energy programme in India, urged for the development of a nuclear power programme . He strongly believed that this would help India achieve a standard of living akin to that of the Western countries.

In addition to the need to meet the rising electricity demand, the drivers for the development of nuclear power also included factors such as the need for diversification of electricity sources to minimise vulnerabilities and to combat the problem of climate change. All the factors have cumulatively boosted the case for nuclear power generation in the country.

Despite the strong political backing across party lines, the development of the nuclear sector in the country has been slow. This has largely been due to the international embargoes placed on India after it conducted the peaceful nuclear test in 1974 and nuclear weapons tests in 1998. Although several delays were caused and the pace of the programme was slowed, the restriction spurred indigenous growth of nuclear technology, as India began to develop alternatives to the denied components and technical expertise.

A few years into the new millennium, avenues to reengage in international nuclear commerce opened up for India. This policy shift was enabled by several factors. Primarily, India had proven itself as a responsible nuclear state as it adhered to a self-moratorium on conducting nuclear tests and observed the principles of the Treaty on Non-Proliferation of Nuclear Weapons (NPT) despite being a non-signatory to it. India's vast market opportunity was also recognised by the West. Furthermore, as debates on sustainable development began to gain traction around the world, India's growing energy requirements and the importance to meet these in a sustainable manner were acknowledged by the US.[2] Other significant geopolitical developments of the time also favoured America's engagement with India. Notable amongst these developments was the revelation of Pakistan's irresponsible behaviour vis-à-vis AQ Khan's proliferation network which led to America's disassociation with Pakistan . In addition, the rise of China featured as an important development that warranted the attention of the West.[3] All these factors created favourable conditions for US engagement with India. Thus, after being excluded from international nuclear commerce for 34 years, India began to discuss its inclusion in it. After a long period of negotiations, primarily between India and the US, the US-India Civil Nuclear Agreement was signed in 2008. The major contours of the agreement were the lifting of sanctions on India, creation of an India-specific waiver by the Nuclear Suppliers' Group (NSG), signing of an additional protocol with the International Atomic Energy Agency (IAEA) by India and the amendment of the US Nuclear Non-Proliferation Act by the US government.[4] On its part, India also committed to a unilateral moratorium on nuclear testing, separation of its civilian and military nuclear facilities and placement of some civil nuclear facilities under IAEA safeguards. Soon after, India signed nuclear cooperation agreements with over 10 countries. Unsurprisingly, India's nuclear programme began to develop at a faster pace. The nuclear reactors whose capacity factor had reduced to below 60 per cent at the beginning of the millennium, began to rapidly increase, reaching close to 80 per cent.

Presently, India has 23 operable reactors contributing 3.1 per cent to the total energy mix of the country and has seven more reactors under

2. For more on this, see Manpreet Sethi, "15 Years of Indo-US Nuclear Deal: Benefits Transcend Nuclear", *Sunday Guardian Live*, July 4, 2020. https://www.sundayguardianlive.com/opinion/15-years-indo-us-nuclear-deal-benefits-transcend-nuclear. Accessed on May 3, 2021.
3. Ibid.
4. Ibid.

construction.[5] While these numbers are below the goals envisaged by the Department of Atomic Energy (DAE) to produce 20 Gigawatts (GW) of nuclear power by 2020, state support to the nuclear sector continues to be strong. India has set new, ambitious goals with a target to increase the share of nuclear power in electricity generation to 9 per cent by the year 2032 and 25 per cent by the year 2050.[6] In May 2021, *Électricité de France* (EDF), filed an offer to supply engineering equipment to build six, third-generation European Pressurised Reactors (EPRs) in Jaitapur. On completion, this will be the world's biggest nuclear power plant and is projected to provide 10 GW of electricity (which is estimated to be able to power around 70 million households) and create about 25,000 jobs during the construction phase, and around 2,700 jobs permanently.[7]

Research on nuclear technology has also been developing steadily. A hallmark of India's civil nuclear strategy is the three-stage programme. Bhabha had stated before the nuclear bombings in Japan, that "when nuclear energy has been successfully applied for power production in, say, a couple of decades from now, India will not have to look abroad for its experts but will find them ready at hand".[8] This has been the endeavour of India's nuclear programme till date. This pursuit for self-sufficiency was envisaged taking into consideration the dearth of uranium resources in India, which works as a fuel in nuclear reactors. Alternatively, it focussed on tapping the country's vast thorium resources instead, which

5. M Ramesh, "Why Nuclear When India has an 'Ocean' of Energy", *The Hindu*, June 13, 2019. https://www.thehindu.com/business/Industry/why-nuclear-when-india-has-an-ocean-of-energy/article28230036.ece. Accessed on April 15, 2021.
6. Niharika Tagotra, "India's Ambitious Nuclear Power Plan—And What's Getting in Its Way", *The Diplomat,* September 9, 2020. https://thediplomat.com/2020/09/indias-ambitious-nuclear-power-plan-and-whats-getting-in-its-way/. Accessed on May 3, 2021.
7. "India Closer to Building the World's Largest Nuclear Plant: French Firm", *Livemint,* April 24, 2021. https://www.livemint.com/news/india/india-closer-to-building-world-s-biggest-nuclear-plant-french-firm-11619242815330.html. Accessed on June 7, 2021.
8. Dr. H.N Sethna. "India's Atomic Energy Programme: Past and Future", *IAEA Bulletin*, Vol. 21, No. 5.

could work as a fuel if immersed in an external supply of neutrons. The three-stage programme was designed as a closed fuel cycle in which each stage can feed into each other, so that the spent fuel produced in the first stage may be reprocessed and used in the second stage to produce power. In the final stage, a sustainable nuclear fuel cycle may be achieved by using thorium.[9] The current generation of nuclear scientists in India continue to remain committed to Bhabha's vision of accomplishing a three-stage cycle.

## Assessing the Challenges in India's Civil Nuclear Programme

Despite coming a long way, the nuclear industry in India continues to be saddled by challenges. Some of the major challenges bedevilling the Indian nuclear industry are the following:

### *1. Public Mistrust in Nuclear Power and Anti-Nuclear Protests*

India has witnessed several anti-nuclear protests over the years that have extended from hunger strikes to demonstrations. These protests have emerged for multiple reasons ranging from land acquisition and rehabilitation to concerns of environmental degradation; from indictment of corruption to concerns about the effect of nuclear power generation on marine life. After the 2011 nuclear disaster in Japan, concerns of nuclear safety once again surfaced, especially among the residents of the Idinthakarai village in Tamil Nadu which had been hit by a tsunami in 2004 and has a nuclear power plant in the region. More recently, in 2019, a cyber attack on the Kudankulam nuclear power plant raised concerns regarding the safety of the reactors.[10] In addition, the nuclear power programme has been mired in other challenges. For example, the development of nuclear power plants in India has been slow in the past few years and has taken several more

9. Tagotra, n. 6.
10. Aathira Konikkara , "There's no Transparency: SP Udayakumar on the Cyber Attack on Kudankulam Nuclear Power Plant", *The Caravan,* November 15, 2019. https://caravanmagazine.in/security/no-transparency-udayakumar-kudankulam-nuclear-plant. Accessed on May 29, 2021.

years than the original deadlines to be completed.[11] Projects have also gone significantly over budget. These factors have exacerbated the anti-nuclear movement in India, which claims that the risks are not worth the underwhelming output.

The protests have significantly delayed the pace of construction and development of nuclear power projects in India. For example the start-up of two Russian Water-Water Energy Reactor (VVERs) was stalled.[12] Public objections even led to the shifting of Westinghouse's first planned project from Gujarat to Andhra Pradesh. In the case of the Kudankulam nuclear power plant, local protests resulted in the halt of commissioning work for nine months.[13]

In addition, the Department of Atomic Energy (DAE) in India has historically been shrouded in secrecy, cosseted by the Atomic Energy Acts of 1948 and 1963 and the Official Secrets Act of 1923. Furthermore, the government's engagement with the public has been limited in the past. The resulting lack of transparency and communication has led to creating vacuums of scepticism and paranoia about nuclear power among the local populace.

However, in the recent past, the Indian government has recognised that the declining credibility of nuclear institutions in India needs to be addressed. Accordingly, the government has taken significant steps to ramp up public engagement and increase transparency. For example , the website of the Atomic Energy Regulatory Board (AERB), in addition to publishing information about the nuclear plants in the country, also includes sections that link the viewer to express any grievances. It also

11. "India's Nuclear Solution to Global Warming is Generating Huge Domestic Protests", publicintergrity.org, December 15, 2015. https://publicintegrity.org/national-security/indias-nuclear-solution-to-global-warming-is-generating-huge-domestic-protests/. Accessed on May 29, 2021.
12. Dan Yurman, "Protests Delay India's Nuclear Renaissance", Ans.org, November 17, 2011, https://www.ans.org/news/article-860/protests-delay-indias-nuclear-renaissance/. Accessed on May 29, 2021.
13. "Indian Government Takes Steps to Get Nuclear Back on Track" , *World Nuclear News,* February 11, 2019. https://www.world-nuclear-news.org/Articles/Indian-government-takes-steps-to-get-nuclear-back. Accessed on May 29, 2021.

provides information with regard to accessing the Right To Information (RTI) section on nuclear power. It also brings out annual reports , bulletins and press releases on issues related to radiological safety, regulatory decisions, effluent discharges and other sensitive issues.[14] The AERB also participates in science and technology fairs and conducts exhibitions to expand public outreach.

*2. Project Delays and Non-Fulfilment of Targets*

In the context of reactors performance, while units such as the ones at the Kaiga power plant have a capacity factor of about 90 per cent, and other plants have crossed 100 per cent, the newer reactors, have been facing challenges in their performances.[15] The Kudankulam Nuclear Power Plant (KKNPP), for example, which began operations in 2013 suffered maintenance issues in 2018 and Unit 1 remained idle for six months.[16] The annual World Nuclear Industry Status Report (WNISR), released in September 2019, revealed that Unit 1 at KKNPP had a load factor of just 54 per cent in 2018 and Unit 2 of 35.2 per cent.[17] This had created some challenges for TANGEDCO, Tamil Nadu's electricity distribution company.[18] In February 2020, the first unit of KKNPP had come to a standstill due to a technical snag. The second unit too

14. Atomic Energy Regulation Board home page, AERB.gov.in. https://www.aerb.gov.in/images/PDF/Annual_Bulletin/2018.pdf. p. 21. Accessed on May 29, 2021.
15. B Sivakumar, "Both Kudankulam Units Shut, 1,000 MW Shortfall for Tamilnadu", *The Times of India,* February 14, 2020. http://timesofindia.indiatimes.com/articleshow/74125162.cms?utm_source=contentofinterest&utm_medium=text&utm_campaign=cppst. Accessed on May 29, 2021.
16. "Kudankulam Reactor Synchronised with Grid", *The Hindu,* May 19, 2019. https://www.thehindu.com/news/national/tamil-nadu/kudankulam-reactor-synchronised-with-grid/article27175281.ece. Accessed on May 29, 2021.
17. "Global Report on Nuclear Energy Flays Kudankulam Units' Performance", *The New Indian Express,* September 15, 2019. https://www.newindianexpress.com/states/tamil-nadu/2019/sep/25/global-report-on-nuclear-energy-flays-kudankulam-units-performance-2038664.html. Accessed on May 28, 2021.
18. "Why the TN Electricity Distributor is in Conflict with Kudankulam Nuclear Power Plant", The news minute, August 28, 2019, https://www.thenewsminute.com/article/why-tn-electricity-distributor-conflict-kudankulam-nuclear-power-plant-107983. Accessed on May 26, 2021.

was, inactive, as it had not resumed operations since it was halted in December 2019. This resulted in a shortage of 1,000 Megawatts (MW) of power for TANGEDCO.[19] Poovulagin Nanbargal, a volunteer group based in Tamil Nadu which works on environmental issues, collected details about the reactor through the RTI Act. The details revealed that Unit 1 tripped more than 40 times since its commission in 2013. Unit 2, likewise tripped 19 times since March 2016.[20]

In addition, the nuclear power programme in India much like elsewhere in the world, has been encumbered with project delays and cost overruns. For example, over Rs 39,000 crore is being invested for the construction of the third and fourth units at KKNP, costing two times more than the construction of the first and second units.[21] One of the major reasons for the cost overruns are the delays faced by the projects. The Prototype Fast Breeder Reactor (PFBR) project, for example, which was sanctioned in 2003, was authorised at a cost of Rs 3,492 crore and was expected to reach completion within seven years. However, the project faced major time overruns, which led to price overruns. The government revised the price estimate to Rs 5,677 crore in 2012.[22]

Similarly, the KKNPP project witnessed major delays, having taken over 28 years to develop, since it was originally planned in 1988. The major reasons for the delays were the dissolution of the Soviet Union and public opposition. Delays are also witnessed in other stages of development such as in purchases. For example, delays in 100 out of 131 high-value purchase orders that were audited by the Controller and Auditor General (CAG), ranged from one day to almost three years![23] A study conducted by Govind Ranmale and B. E. Narkhede that assessed the causes for delays in Indian nuclear power plants, attributed the problem to factors

19. Sivakumar, n. 15.
20. n. 18.
21. Ibid.
22. "India's First Prototype Fast Breeder Reactor Has a New Deadline: Should We Trust It?", *The Wire,* August 20, 2021. https://science.thewire.in/the-sciences/prototype-fast-breeder-reactor-dae-bhavini-npcil-liquid-sodium-coolant-purchase-orders/. Accessed on May 25, 2021.
23. Ibid.

such as dearth in planning, unavailability of materials, issues of quality assurance and stakeholder management, public protests and delays in payment, among others.[24] The economic status of a country could be a factor too, as it is generally known that developing countries have longer gestation periods.

However, the Indian nuclear establishment has been fairly successful in overcoming the traditional challenges in the recent past. For example, the construction of the Tarapur Atomic Power Plant (TAPP) 3 and 4 was not only done well in time but also cost less than what was originally allocated for the project.[25] In addition, the nuclear institutions too have been coming up with creative problem solving techniques. For example, to avoid land acquisition problems, new reactors are being built in old nuclear parks instead of setting up new ones.

## Assessing the Future Trajectory of India's Civil Nuclear Power Programme

The future of India's nuclear power programme is likely to be influenced by the following factors.

### *1. Ambitious Targets*

In 2018, Kaiga 1, an indigenously developed Pressurised Heavy Water Reactor (PHWR), in the Indian state of Karnataka, set a new world record for operating incessantly for 962 days from May 13, 2016 till December 31, 2018.[26] At the time, the plant load factor was over 99 per cent.[27] Similarly, India's Rajasthan 3 and 5 reactors continuously

---

24. Govind Ranmale and B. E. Narkhede, "Delays in Indian Nuclear Power Plant", Researchgate, January 2015. https://www.researchgate.net/publication/300702662_Delays_in_Indian_Nuclear_Power_Plant. Accessed on June 2, 2021.
25. Dr. Manpreet Sethi, "The Nuclear Energy Imperative: Addressing Energy Poverty, Energy Security and Climate Change in India", *India Infrastructure Report 2010,*
26. The record was beaten by Canada's Darlington Unit 1 in February 2021.
27. "Indian-Designed Nuclear Reactor Breaks Record for Continuous Operation", powermag.com, February 1, 2019. https://www.powermag.com/indian-designed-nuclear-reactor-breaks-record-for-continuous-operation/. Accessed on May 30, 2021.

operated for 777 and 765 days respectively.[28] Despite achieving such significant milestones, the overall developmental trajectory of India's nuclear power programme has been underwhelming, marked by delays and slow growth. With a civil nuclear programme having been set up over six decades ago, nuclear power contributes to less than 3 per cent of India's current electricity generation, having an installed capacity of 6,885 Megawatts electric (MWe).[29] This figure is much less than the 8,000 MW of nuclear power capacity that Homi Bhabha had envisioned for the year 1980, back in 1954.[30] While the international sanctions on India that lasted for over three decades were one of the key factors responsible for the slow growth, a major expansion in the programme was expected after the 2008 India- US nuclear deal. However, several other factors continue to limit the growth of the programme such as anti-nuclear protests and delays in the development of the second stage of the nuclear fuel programme, among others.[31]

The near-term growth rate of the nuclear power sector in the country appears to be slow, as the targets have not been met. For example, India set a goal of generating 20,000 MW of electricity through nuclear power by the year 2020. However, the nuclear capacity in 2020 was only 6,780 MW.[32] In 2011 again, India set a target of achieving up to 14,600 MWe by the year 2032. However, the chances of missing the new targets seem likely as the new projects set up to achieve these goals are way behind schedule.[33]

The government, however, is committed to developing the country's nuclear power programme. The Department of Atomic Energy is aiming

---

28. Ibid.
29. "Nuclear Power in India", World-nuclear.org. https://world-nuclear.org/information-library/country-profiles/countries-g-n/india.aspx. Accessed on July 15, 2021.
30. Prerna Gupta, M V Ramana, "A Decade After the Nuclear Deal", *The India Forum*, May 26, 2019. https://www.theindiaforum.in/article/decade-after-nuclear-deal . Accessed on July 17, 2021.
31. n. 13.
32. "Nuclear Power Capacity to Treble by 2031: Govt", *The Tribune*, February 15, 2021. https://www.tribuneindia.com/news/nation/nuclear-power-capacity-to-treble-by-2031-govt-212513. Accessed on June 3, 2021.
33. Ranmale and Narkhede, n. 24.

to increase the nuclear power generation capacity over six fold in the coming decade.[34] Several steps have been taken to achieve this goal. For example, in 2017, over Rs 1 lakh crore was approved for the setting up of nuclear power plants. Scheduled to be completed by the year 2031, these are expected to have a combined capacity of 7,000 MW.[35] In 2020, Minister of State for the DAE, Jitendra Singh, stated that to augment nuclear power production, nine new reactors are expected to achieve completion by 2024-25, in addition to 12 other reactors which were given the government's approval in 2017.[36] Other sites too have gained 'in principle' approval by the authorities for setting up more reactors in the future.[37] In this regard, the DAE, has stated that the nuclear power capacity of India will increase from the current 6,780 MW to 22,480 MW by the year 2031.[38] To intensify near-term growth, several uranium mining operations too have been initiated such as the Rohil Uranium Project in Rajasthan, Kanchankayi Uranium Project in Karnataka, Jajawal Uranium Project in Chhattisgarh and Chitrial Uranium Project in Telangana, among others.

### *2. Strides in Nuclear Research and Development*

The Indian nuclear industry has made significant strides in the field of nuclear research and development. In 2012, Siegfried Hecker, a former director of the Los Alamos National Laboratory in the US, stated in an issue of *Physics Today* that "India has the most technically ambitious and innovative nuclear energy program in the world…the extent and

34. Mohana Basu, "Govt Sets Target to Increase Nuclear Power Generation Capacity by Over 3 Times in 10 Years", *The Print,* March 4, 2020. https://theprint.in/india/govt-sets-target-to-increase-nuclear-power-generation-capacity-by-over-3-times-in-10-years/375366/. Accessed on June 1, 2021.
35. "Myths Linked to Nuclear Energy Should be Busted: MoS Jitendra Singh", *Financial Express,* October 19, 2019. https://www.financialexpress.com/industry/myths-linked-to-nuclear-energy-should-be-busted-mos-jitendra-singh/1740042/. Accessed on June 1, 2021.
36. n. 13.
37. Ibid.
38. n. 32.

functionality of its nuclear experimental facilities are matched only by those in Russia and are far ahead of what is left in the US."[39] Most notable among India's scientific endeavours in the field of nuclear technology is the development of the three-stage cycle. The reactor system is designed to utilise the vast thorium resources in the country. To materialise this process, India today is a world leader in thorium research and had published double the number of research papers than its competitors between the years 2002 to 2006.[40] The United Kingdom too, has been collaborating with India to take the research forward. Five research proposals of £2 million are being funded by the UK's Engineering and Physical Sciences Research Council and by India's Department of Atomic Energy.[41]

India's nuclear power programme was formulated to be a closed fuel cycle one that could be achieved in three successive stages. In the first stage, natural uranium is used as fuel in a pressurised heavy water reactor. The by-product of it, plutonium-239 is then to be used to make mixed oxide (MOX) fuels for the Fast Breeder Reactor (FBR) in the second stage. India has successfully completed the first stage of the nuclear fuel programme and the second stage is very close to achieving success . For this purpose, India has been operating a Fast Breeder Test Reactor (FBTR) since 1985 to develop a future breeder reactor for the second stage. India, currently is at the verge of launching a Prototype Fast Breeder Reactor (PFBR). Located in Tamil Nadu, it has a capacity of 500 MW. The completion of the PFBR has been delayed over the years, but it is expected to finally get commercialised in October 2022. Tracing the development of the reactor, an official stated during the Rajya Sabha question and answer session that "all construction activities

39. S Raghotham, "What Got Made in India Before Modi, How it Was Done", Indiadefensenews, December 22, 2019. http://www.indiandefensenews.in/2019/12/what-got-made-in-india-before-modi-how.html. Accessed on June 12, 2021.

40. Srikumar Banerjee, "Thorium Utilisation for Sustainable Supply of Nuclear Energy," *Virginia Tech,* September 2010. Accessed on June 2, 2021.

41. "A Future Energy Giant? India's Thorium-Based Nuclear Plans", Physics.org, October 1, 2010. https://phys.org/news/2010-10-future-energy-giant-india-thorium-based.html. Accessed on June 2, 2021.

(have been) completed and integrated commissioning of all systems (is) in progress."[42] The standing committee in this regard has said, "Even though it would have taken almost two decades (of work on the PFBR) when commissioning takes place, this is a pioneering initiative of which India can be justifiably proud (*sic*). It will transform our nuclear energy programme."[43] Plans are also in progress to build 5 more fast breeder reactors of a combined capacity of 2.5 GW (Gigawatts). Research on the final stage, to develop thorium-based reactors is also in full swing and is expected to be successful in the next few decades. Once proven technologically, this would be a major breakthrough in the field of nuclear power generation, as it not only minimises nuclear waste but is also more proliferation resistant. The pace at which India develops its three-stage cycle will be a major determinant of the future of nuclear power in the country. This would help India attain energy independence as experts estimate that through this process, India will be able to produce about 500 GWe of power for at least 400 years using the economically extractable thorium reserves in the country.[44] This could also help in launching India as a leading exporter of alternate nuclear technology.

India has also been exploring the prospects of nuclear fusion technology. It is one of the seven countries funding the International Thermonuclear Experimental Reactor (ITER) project in France.[45] The ITER project is working on building the world's largest nuclear fusion reactor, which can generate ten times more energy than it consumes. Larsen and Toubro, an Indian construction engineering company, has also developed a component for this project.

---

42. n. 22.
43. Ibid.
44. T.S. Subramanian, "A Debate Over Breeder Reactors", *Frontline,* Vol. 15, No. 25, December 1998. Accessed on June 2, 2021.
45. Karan Kamble, "Not On The Money—India's Fusion Energy Ambitions May Be Hampered By Understaffing at The World's Largest Fusion Reactor", swarajyamag.com, August 28, 2020. https://swarajyamag.com/science/not-on-the-money-indias-fusion-energy-ambitions-may-be-hampered-by-understaffing-at-iter. Accessed on June 3, 2021.

### *3. Nuclear Exports and Services*

As India has successfully developed nuclear power knowhow, it is now moving beyond just domestic development of nuclear power. Apart from extending training and capacity building assistance to other countries, it is now poised to export nuclear technology as well. However, since India is not a member of the NSG, exporting nuclear technology becomes complicated for the country. It has, thus, reached an agreement with Russia, allowing it to circumvent this problem. In 2015, India signed a 'Strategic Vision for Strengthening Cooperation in Peaceful Uses of Atomic Energy' with Russia. Under this agreement the "two sides will explore opportunities for sourcing materials, equipment and services from Indian industries for the construction of the Russian-designed nuclear power plants in third countries".[46]

In extension, India signed a tripartite agreement with Russia and Bangladesh in 2018 to develop the Rooppur Nuclear Power Plant project in Bangladesh. India's contribution to the Rooppur project will be through the construction of non-critical infrastructure, capacity building and installation work, in addition to supplying materials and equipment.[47] India's Hindustan Construction Company has engaged in a joint venture with the Bangladesh-based MAX Group as well. In addition, India has extended its training services to Bangladeshi scientists at its Kudankulam nuclear plant, which too was built with Russian assistance. Bangladesh President Sheikh Hasina also signed three agreements to further nuclear cooperation with India during her visit to the country. In addition, the Bangladesh Atomic Energy Commission (BAEC) and India's Global Centre for Nuclear Energy Partnership (GCNEP) also took the cooperation ahead, with the appointment of GCNEP as the

---

46. "Strategic Vision for Strengthening Cooperation in Peaceful Uses of Atomic Energy Between the Republic of India and the Russian Federation", Ministry of External Affairs, Government of India. https://mea.gov.in/bilateral-documents.htm?dtl/24487/Strategic_. Accessed on April 5, 2021.
47. "Russia and India Sign Action Plan on New Projects", *World Nuclear News*, October 5, 2018. https://www.world-nuclear-news.org/Articles/Russia-and-India-sign-action-plan-on-new-projects. Accessed on April 5, 2021.

consultant for the construction and operation of the Rooppur project.[48] In March 2021, it was announced that Indian companies will be assisting Bangladesh in developing transmission lines for the Rooppur plant, which are estimated to be valued at more than US$ 1 billion.[49] This provides India an opportunity to prove its mettle as a nuclear exporter. Establishing itself in the nuclear exports market won't just advance its domestic "Make in India" initiative but would also help in the promotion of nuclear power as a foreign policy tool.

## Conclusion

Several pressing issues demand boosting nuclear power production in India. First, India is expected to soon overtake the European Union and become the world's third largest consumer of energy. Second, India is presently the third largest emitter of $CO_2$ in the world. In addition, energy imports are mounting and industrial growth is being challenged by the power crises. These factors compel India to promote a sustainable and clean source of power to spur economic growth, increase the standard of living of its people and alleviate the climate change crisis. In this regard, a mixed energy basket with a significant role for nuclear power could go a long way in powering the growth of the country and minimising many of the challenges facing Indian society.

India's future as a potential nuclear exporter looks promising. With a long history of research and development in the field of nuclear technology, India appears to be ready to cash in on its technical prowess. Currently, the nuclear exports market is dominated by Russia and China, both of which have been tainted with allegations of corruption and political manipulation. In this regard, with its history of benign foreign policies, India could emerge as a favourable country to engage in nuclear

48. "India's Credit Line For Bangladesh Covers Nuclear Projects, Rooppur Nuclear Power Plant: Foreign Secretary", NDTV.com., March 28, 2021. https://www.ndtv.com/india-news/indias-credit-line-for-bangladesh-covers-nuclear-projects-rooppur-nuclear-power-plant-foreign-secretary-2400332. Accessed on April 5, 2021.
49. Ibid.

exports. Nuclear exports, would not only be economically beneficial for India but also be a strategic foreign policy tool.

However, to export nuclear technology, it is important for India to accelerate its own domestic programme. There is, thus, a need for a comprehensive energy strategy, with a clear roadmap and thorough implementation. If price overruns and prolonged delays of projects continue to plague the nuclear programme, the Indian nuclear industry will be at risk of collapse. A more proactive and robust approach is needed to move out of the slump. Thus, the future of India's energy security and its nuclear power sector depend on how India responds to the many challenges it faces.

Most pertinent among these challenges that requires immediate attention is that of public opposition. Increased political support and boosting finances will not be sufficient if the mood of the nation is anti-nuclear. Public support is paramount for the survival and success of a nuclear power project in a democratic country. Gaining credibility could go a long way in consolidating public support. There is also a need for an energised public communication strategy which can generate a better understanding of nuclear power to its citizens. Especially in areas selected to host nuclear projects, it is vital to proactively engage with the local population, communicate about the various benefits they offer, and address the concerns or grievances of the people. It would be unfortunate if after winning international acceptance, India loses domestic support.

# 3. Emergence of Small Modular Reactors: Challenges and Opportunities

*Manpreet Sethi and Zoya Akhter*

In 1954, when Lewis L. Strauss, then chairman of the US Atomic Energy Commission, pronounced that nuclear energy would soon become "too cheap to meter", it ushered in a wave of nuclear optimism around the world. Unfortunately though, the civil nuclear industry was never able to deliver on this promise. Factors such as high cost and long gestation of reactor construction, limitations of grid capacity and land availability, as well as proliferation concerns, constantly weighed down the nuclear industry. Nuclear accidents in Three Mile Island in 1979, Chernobyl in 1984, and Fukushima in 2011 further dented public trust in nuclear safety.

In the 2020s, however, a new nuclear enthusiasm is being rekindled by the idea of "Small Modular Reactors" (SMRs). These are being proffered as a solution to some of the challenges associated with traditional nuclear reactors. A race is on within the nuclear industry to proffer new designs and manufacturing processes and seek their certification by national regulatory agencies, as well as the International Atomic Energy Agency (IAEA). These efforts are receiving considerable government backing in the hope that they will prove their commercial viability and become operational within this decade. The United States Department of Energy (DoE), for example, is providing cost-sharing support to select SMR companies via public-private partnerships and granting them access to experimental facilities housed at national laboratories. The United Kingdom also is providing financial support to SMRs as part of the technology portfolio to reach its 2050 carbon neutrality objective.[1] Some estimates project that nearly 21 Gigawatts (GW) of electricity may be globally generated by SMRs by 2035.[2] However, issues of economic feasibility, liability and security of the SMRs remain a concern. Most designs are presently at the stage of being tested as a prototype or as a first of a kind demonstration.

This paper attempts to understand the excitement around SMRs. It tracks the ongoing developments in this field across countries. It examines the rationale for such reactors, assesses the benefits they claim, and the challenges that remain. It concludes by exploring the role SMRs could play in India's energy strategy.

## Concept and Definition of SMRs

Two main attributes are being associated with SMRs—small and modular. The World Nuclear Association has defined SMRs as "nuclear reactors generally 300 Megawatts electric (MWe) equivalent or less, designed with modular technology using module factory fabrication, pursuing economies of series production and short construction times."[3] Some countries, like the UK, however, have found this definition too narrow to accommodate their design of Natrium technology, which has a slightly higher capacity of 345 MW.[4] So, the UK prefers to use the term "Advanced Nuclear Technologies."[5]

1. OECD/NEA," Small Modular Reactors: Challenges and Opportunities", *Nuclear Development, OECD*, Nuclear Publishing, Paris, 2021. https://doi.org/10.1787/18fbb76c-en.
2. "Small Modular Reactors: The Dawn of a New Nuclear Era", *Financial Express*, August 4, 2021. https://www.financialexpress.com/defence/small-modular-reactors-the-dawn-of-a-new-nuclear-era/2304153/. Accessed on October 1, 2021.
3. "Small Nuclear Power Reactors", world-nuclear.org. https://www.world-nuclear.org/information-library/nuclear-fuel-cycle/nuclear-power-reactors/small-nuclear-power-reactors.aspx. Accessed on September 2, 2021.
4. Arjun Makhijani and M. V. Ramana, "Can Small Modular Reactors Help Mitigate Climate Change?", The Bulletin.org, July 21, 2021. https://thebulletin.org/premium/2021-07/can-small-modular-reactors-help-mitigate-climate-change/. Accessed on October 3, 2021.
5. "What is a Small Modular Reactor (SMR)? (A Complete Guide)", Twi-global.com. https://www.twi-global.com/technical-knowledge/faqs/small-modular-reactor. Accessed on September 8, 2021.

SMRs may broadly be understood as nuclear fission reactors with a capacity of about 300 MWe or less in comparison to the traditional reactors of 1,000 to 1,600 MW capacity. It is assumed that they would be capable of being centrally manufactured at a factory and then transported to the desired site for assembly/installation. Owing to this characteristic, SMRs have been described as a product and not a project. Much like a machine, the reactor would arrive at a site, be plugged in and start producing electricity. SMRs would also offer the possibility of addition of multiple similar reactors when desired. Also, these could be placed on land, on a ship for offshore deployment, or even in an underground or submerged environment.

While the idea of modularity may be new, the idea of small reactors is not. It can be traced back to the 1940s when the US military had initiated research programmes to develop smaller nuclear reactors. The first such civilian small reactor was commissioned in 1955 in Minnesota. But, it faced major problems of cost overruns, exceeding its budget by $9.8 million. Safety concerns too arose as cracks developed in its cooling system just years after its commissioning.

Thereafter, the inclination shifted towards larger reactors that could exploit economies of scale and effort. Driven by a desire to maximise megawatts and, thus, amortise the Operating and Maintenance (O&M) costs, the nuclear industry focussed on larger capacity reactors in the range of 1,000-1,700 MWe. However, some countries continued to make small reactors.[6] For example, India's 220 MWe Pressurised Heavy Water Reactors (PHWRs) can be seen to fall in this category in terms of capacity, though not from the point of view of modularity of construction.

Modular construction of the smaller variety of reactors has caught the attention of the nuclear industry more recently. Accordingly, over 70 SMR designs are being developed across the world today. At different

6. For more on this, see Lois Parshley, "The Countries Building Miniature Nuclear Reactors", *BBC*, March 9, 2020. https://www.bbc.com/future/article/20200309-are-small-nuclear-power-plants-safe-and-efficient. Accessed on September 12, 2021.

stages on the drawing board, these designs range from being slightly modified versions of existing reactors to those involving completely new technologies. Design types include thermal-neutron, fast-neutron, gas-cooled, light-water, graphite-moderated high temperature and generation IV reactors. In addition, there are also land-based and marine-based reactor designs, envisaging utilisation of different fuel and coolant options. The SMR landscape, therefore, is abuzz with activity as new concepts are being developed, and also as nuclear companies restructure, abandon, or combine efforts.

Given this level of activity, the IAEA too has set up the SMR Regulators' Forum to help countries navigate the challenges and enable sharing of information on issues of common concern. In fact, the IAEA published a Technology Roadmap for Small Modular Reactor Deployment in 2021 that identifies, evaluates and promotes collaboration and knowledge sharing amongst technology developers, industry, users and regulatory bodies towards a common objective. The following section scans the major developments across countries. These are presented in the order of level of activity in the SMR field.

## National SMR Efforts

### *Russia*

Russia is a frontrunner in SMRs, being the only country to have already made such a reactor operational. This feat was especially remarkable since it also became the first Floating Nuclear Power Plant (FNPP). In December 2019, two units of the KLT-40S, installed on a barge, the *Akademik Lomonosov,* were connected to the electricity grid. Commercial operation commenced in May 2020.[7] Docked at Pevek, the two reactors, producing 35 MWe each, replaced a coal-powered

7. "Russia Connects Floating Plant to Grid", *World Nuclear News,* December 19, 2019. https://world-nuclear-news.org/Articles/Russia-connects-floating-plant-to-grid#:~:text=The%20floating%20nuclear%20power%20plant,Pevek%2C%20in%20Russia's%20Far%20East. Accessed on October 2, 2021.

plant in the remote town and are estimated to generate electricity for about 200,000 people.[8]

However, it needs to be noted that it was in 2003 that the Russian Federation's nuclear regulator, Rostechnadzor, had issued the first construction licence for the *Akademik Lomonosov.* Construction of the first domestic design of this unit began in 2007 and was completed in the beginning of 2018. The application for the operating licence for the plant was then submitted to Rostechnadzor. After initial testing in St. Petersburg, the barge was towed to Murmansk, where nuclear fuel was loaded and the final tests conducted, before it was finally towed to, and deployed in, Pevek.

From the above timelines, it is clear that the time taken for the first FNPP, from receiving the construction licence to becoming operational, was not as short as has been anticipated for smaller reactors. The cost of the reactor also turned out to be four times more than the original estimate.[9] While this may be attributed to the novelty of the technology, it is equally true that Russia has long experience in nuclear construction, including on ships and submarines. Whether the gestation period of SMRs can improve in the future will become clear as Russia develops and deploys more SMRs, one of which is due to be commissioned in 2027.[10] For Russia, a particular reason for developing such plants is the need to provide power to its far-flung regions. Accordingly, in 2021, Russia signed an agreement for the construction of four SMRs to provide electricity to the Baimskaya region which is rich in copper and gold reserves, as also to the North Yakutsk Arctic zone.[11]

---

8. "Akademik Lomonosov Floating Nuclear Co-Generation Plant", *Power Technology*, https://www.power-technology.com/projects/akademik-lomonosov-nuclear-co-generation-russia/. Accessed on October 2, 2021.
9. Charles Digges, "New Document Show Cost of Russia Feating Nuclear Power Plant Skysckets", *Bellona*, May 25, 2015.
10. "Six Russian SMR Designs", *NEI Magazine,* January 16, 2019, https://www.neimagazine.com/features/featuresix-russian-smr-designs-6939130/. Accessed on October 3, 2021.
11. "SMRs to Power Arctic Development", *World Nuclear News,* September 3, 2021, https://world-nuclear-news.org/Articles/SMRs-to-power-Arctic-development. Accessed on October 18, 2021.

### *China*

China too has several SMR designs in the pipeline. In July 2021, it began construction of its first indigenously developed, commercial onshore nuclear project.[12] The "Linglong One", also known as ACP-100, is a multi-purpose pressurised water reactor. Interestingly, it is also the first SMR design to pass a safety review of the IAEA in 2016.[13] Once completed, in about 58 months, it is projected to meet the energy demands of 526,000 households.[14]

Another prototype, the High Temperature Gas Cooled Reactor–Pebble Bed Module (HTRPM) was connected to the grid in February 2022 in Shandong province.[15] Construction of this reactor began in December 2012, with two reactor units planned of 200 MWe. In addition, China is developing other designs such as the CAP200, DHR400 and HAPPY200. With its focus on nuclear Research and Development (R&D), China is likely to dominate the SMR industry along with Russia. A report released by Wood Mackenzie in this regard projected that China will likely own the largest share of the global pipeline for SMRs by 2050.[16]

### *Argentina*

Argentina has been developing an SMR called the Central Argentina de Elementos Modulares (CAREM) reactor. Construction of the prototype, an integral type[17] pressurised Light Water Reactor (LWR)

---

12. n. 10.
13. Pamela Largue, "China Nuclear Starts Building Small Modular Reactor Demo Project", *Power Engineering,* July 14, 2021. https://www.powerengineeringint.com/nuclear/china-nuclear-starts-building-small-modular-reactor-demo-project/. Accessed on October 12, 2021.
14. Ibid.
15. "China Makes World's First Small Modular Reactor", *Bloomberg*, February 4, 2022.
16. "China to Lead Small Modular Reactor Market by 2050—Report", *Power Engineering International,* August 30, 2021. https://www.powerengineeringint.com/nuclear/china-to-lead-small-modular-reactor-market-by-2050-report/. Accessed on October 23, 2021.
17. All the traditional primary coolant system components have been incorporated inside the reactor vessel (i.e. integral to the reactor vessel).

with a rated electrical power of 27 MWe began in 2012. Its due date for commissioning is late 2023, which looks ambitious since the project has been facing several political and economic hurdles.[18] It was only in July 2021 that Argentina's national atomic energy commission, the CNEA, and the state-owned nuclear plant operator Nucleoeléctrica Argentina signed a deal to take forward the construction of the SMR. The regulator has established a staged licensing approach aimed at clearing every next step only after successful demonstration of specific safety milestones. Once ready, the operating experience so gained will be used to support a future larger version planned of between 150 and 300 MWe.[19] Such reactors are being designed for places with small electricity grids, as also desalination plants.

### *USA*

The United States government too has been keen on developing SMR technology. The US Department of Energy (DoE) has acknowledged such reactors as a "key part of the department's goal to develop safe, clean, and affordable nuclear power options."[20] Currently, the technology appears to have bipartisan support. Under President Trump, the DoE approved a $1.4 billion grant towards development of this technology.[21] Likewise, the present US President Joe Biden too has recently announced the plan to launch the Foundational Infrastructure for Responsible Use of Small Modular Reactor Technology (FIRST)

18. "Techint Halts Work on Argentina's Pioneering Small Modular Reactor Scheme", *Global Construction Review,* November 15, 2019. https://www.globalconstructionreview.com/techint-halts-work-argentinas-pioneering-small-mod/, accessed on September 13, 2021.
19. "Technology Roadmap for Small Modular Reactor Deployment", *IAEA Nuclear Energy Series,* No. NRT1.18.
20. "Advanced Small Modular Reactors (SMRs)", Energy.gov. https://www.energy.gov/ne/advanced-small-modular-reactors-smrs. Accessed on October 26, 2021.
21. Caroline Delbert, "Trump Gives Tiny Nuclear Reactors a Billion-Dollar Boost. It's a Huge Win for the Startup Leading the Nuclear Revolution", *Popular Mechanics,* October 21, 2020. https://www.popularmechanics.com/science/a34431835/nuscale-small-modular-reactors-government-grant/. Accessed on October 23, 2021.

programme. It is tasked to provide capacity building support to partner countries to jointly develop their nuclear energy programmes. The Department of State has pledged over $5 million as initial investment to support the project. The Biden Administration has endorsed the FIRST programme as one of the primary efforts by the US government to promote innovation in clean energy technologies to combat the climate change crisis.[22]

Even as several designs and capacities from 160MW to 300 MW are being experimented with, one particular design of an LWR is being reviewed for licensing by the Nuclear Regulatory Commission (NRC). The DoE has also partnered with NuScale Power, a private American company and Utah Associated Municipal Power Systems (UAMPS), a political subdivision of the state of Utah, to develop another kind of SMR at the Idaho National Laboratory. NuScale estimates to deliver its 77 MW reactor by 2027 and make it operational by 2029. Thereafter, it plans to sell 11 more reactors by 2030, some to the US Department of Energy and others to UAMPS member utilities.[23] Meanwhile, in January 2020, the BWRX-300 design also announced its entry into the licensing process, with an estimated operationalisation by the end of this decade.[24] In addition, the nuclear reactor company of Bill Gates, "TerraPower", and Warren Buffet's "PacifiCorp" too have recently announced their joint

22. Media Note, "Program To Create Pathways to Safe and Secure Nuclear Energy Included in Biden-Harris Administration's Bold Plans To Address the Climate Crisis", US Department of State, April 27, 2021. https://www.state.gov/program-to-create-pathways-to-safe-and-secure-nuclear-energy-included-in-biden-harris-administrations-bold-plans-to-address-the-climate-crisis/. Accessed on October 26, 2021.
23. Dave Levitan, "First US Small Nuclear Reactor Design Is Approved: Concerns About Costs and Safety Remain, However", *Scientific American,* September 9, 2020. https://www.scientificamerican.com/article/first-u-s-small-nuclear-reactor-design-is-approved/. Accessed on October 2, 2021.
24. "First US Small Modular Boiling Water Reactor Under Development", US Office of Nuclear Energy, February 19, 2020. https://www.energy.gov/ne/articles/first-us-small-modular-boiling-water-reactor-under-development. Accessed on September 14, 2021.

Natrium reactor project to be constructed at Wyoming, at the site of an old coal plant.[25]

*United Kingdom*

Prime Minister Boris Johnson has expressed interest in, and support for, SMRs. With some of the existing nuclear reactors due to retire in 2035, and given the dearth of suitable sites for larger nuclear plants, SMRs have gained traction in the country. The Ten-Point Plan and the Energy White Paper released by the UK government mentions plans to deploy an SMR in the early 2030s.[26] Towards this end, the government has invested £215 million to design such reactors as part of the Low Cost Nuclear (LCN) programme. An additional £170 million for R&D has also been approved.[27] In October 2021, a potential site was identified for the country's first such reactor.[28] The government is also providing financial support to private companies to develop this technology. For instance, the Rolls Royce Consortium is developing an SMR that can be scaled for producing from 220 to 440 MWe. Currently in its early phase, it is estimated to become operational in a decade.[29]

*Canada*

In 2018, Canada decided to facilitate the development of SMRs through a dedicated roadmap in order to actively engage with local,

25. Timothy Gardner and Valerie Volcovici, "Bill Gates' Next Generation Nuclear Reactor to be Built in Wyoming", *Reuters,* June 3, 2021. https://www.reuters.com/business/energy/utility-small-nuclear-reactor-firm-select-wyoming-next-us-site-2021-06-02/. Accessed on October 2, 2021.
26. "Advanced Nuclear Technologies", Department for Business, Energy and Industrial Strategy, Government of UK. https://www.gov.uk/government/publications/advanced-nuclear-technologies/advanced-nuclear-technologies. Accessed on October 27, 2021.
27. Ibid.
28. "Mini Nuclear Reactors Vie for Key Role in UK's Push to Hit Climate Targets", *Financial Times.* https://www.ft.com/content/7da30202-2db9-4ab3-9428-458a9d8728bd. Accesssed on October 15, 2021.
29. n. 5.

national and international stakeholders. In 2019, three Canadian provinces signed a memorandum to collaborate in developing and deploying SMRs. The Canadian Nuclear Research Initiative was also launched as a programme to support collaborative SMR research projects by connecting global vendors of SMR technology with facilities and expertise within Canada's national nuclear laboratories. In the following year, the Canadian government released its SMR Action Plan, expressing a desire to emerge as a global leader in SMR technology. The plan states, "With over 60 years of science and technology innovation, a world-class regulator and a vibrant domestic supply chain, Canada's nuclear industry is poised to be a leader in an emerging global market estimated at $150 billion a year by 2040..."[30]

### *France*

In April 2021, France launched a strategic plan for the French nuclear industry. Of the €500 million that it earmarked for the nuclear industry, €70 million was allocated for the construction of a prototype SMR.[31] Électricité de France (EDF), the French multinational electric utility company, intends to finish designing the reactor by 2022 and then build a pilot project by the end of the decade. In October 2021, President Macron announced another €30 billion for a five-year promotion strategy for the country's nuclear power sector, in which he announced development of "innovative small-scale nuclear reactors" by 2030 as the country's "number one priority".[32]

30. "Canada's Small Modular Reactor Action Plan", Nuclear Energy and Uranium, Government of Canada. https://www.nrcan.gc.ca/our-natural-resources/energy-sources-distribution/nuclear-energy-uranium/canadas-small-nuclear-reactor-action-plan/21183. Accessed on October 6, 2021.
31. Daniel Garton, Richard Hill et al, "Why Small Modular Reactors will Shape the Future of Nuclear Debate", *White & Case*, June 8, 2021. https://www.whitecase.com/publications/insight/why-small-modular-reactors-will-shape-future-nuclear-debate. Accessed on August 7, 2021.
32. "France Unveils Nuclear Power Overhaul—With an Eye on China", France24.com, October 13, 2021. https://www.france24.com/en/france/20211013-france-unveils-nuclear-power-overhaul-with-eye-on-china. Accessed on October 28, 2021.

### *Other Countries*

Several other countries too have been making efforts at developing SMRs. For example, in the Czech Republic, GE Hitachi Nuclear Energy (GEH) and ČEZ, an electricity conglomerate of 96 companies signed a memorandum in 2020 to assess the potential for developing SMRs.[33] Likewise, in northern Europe, in March 2021, Fermi Energia, an Estonian nuclear energy start-up, signed a Memorandum of Understanding (MoU) with Rolls-Royce to study the potential of deploying SMRs as part of the country's low carbon energy strategy. Meanwhile, Ukraine and Poland have signed agreements with the American company NuScale to deploy SMRs in the near future.[34] Similarly, Saudi Arabia has signed a pre-project engineering contract with South Korea for SMR construction.[35] There is also significant interest for SMRs in the Southeast Asian region. The Association of Southeast Asian Nation's (ASEAN's) region-wide energy cooperation plan in this regard discusses the scope of developing SMRs.[36]

As is evident from this brief scan of the global SMR landscape, optimism for the technology is high. Governments and industries are upbeat as the former provide an enabling environment to the latter. The following sections now examine the advantages and the challenges of this new technology.

---

33. "GE Hitachi Nuclear Energy and Čez Announce Small Modular Reactor Technology Collaboration in the Czech Republic", Cez.Cz. February 3, 2021. Https://Www.Cez.Cz/En/Media/Press-Releases/Ge-Hitachi-Nuclear-Energy-And-Cez-Announce-Small-Modular-Reactor-Technology-Collaboration-In-The-Czech-Republic-80848. Accessed on October 28, 2021.
34. "NuScale SMR Under Consideration for Ukraine", *World Nuclear News*, September 2 2021. https://www.world-nuclear-news.org/Articles/NuScale-SMR-under-consideration-for-Ukraine. Accessed on October 18, 2021.
35. "SMR in the Making", *Rosatom Newsletter*, March 2020. https://rosatomnewsletter.com/2020/03/25/smr-in-the-making/. Accessed on October 25, 2021.
36. Miguelle Santini at the CIL Webinar on "A Multi-Pronged Approach towards Small Modular Reactor Regulation", July 28, 2021.

## Advantages of SMRs

### *Speed of Installation and Cost Savings*

The biggest advantage of SMRs has been perceived in their ability to be manufactured as pre-fabricated and pre-tested modules that can be easily assembled/installed on-site after being transported conveniently. Modularity in manufacturing and assembly is expected to reduce construction time by compressing a large technological project into basically installing a pre-fabricated product of standardised quality. A study in this regard has suggested that SMRs can reduce construction time by 3.5 years in comparison to the average 6.5 years required as of now.[37] Reduced gestation periods are then expected to reduce capital costs too and, thus, increase the economic competitiveness of the reactors. This could increase the financial affordability of nuclear power to countries which have traditionally found it too expensive because of the high capital costs involved. Design standardisation and modularisation are expected to enable factory fabrication.

### *Enhanced Safety and Security Features*

Given the new technologies being incorporated by SMRs, these are expected to include several enhanced safety and security features such as lower thermal power of the core, and compact architecture. A smaller core is expected to offer the benefit of reducing the risk of radiation exposure in case of an accident, as well as the need for Emergency Planning Zones (EPZs). Such benefits could mean that some SMRs may be located closer to where energy is needed and, thus, reduce elaborate grid requirements.

SMRs are also expected to incorporate more modern passive safety systems that minimise the need for human intervention in case of emergencies. For example, Westinghouse claims that their SMRs are

37. Clara A Lloyd, " Modular Manufacture and Construction of Small Nuclear Power Generation Systems", *Research Gate,* May 2019. https://www.researchgate.net/publication/337937287_Modular_Manufacture_and_Construction_of_Small_Nuclear_Power_Generation_Systems. Accessed on October 3, 2021.

being designed to enable automatic reactor shutdown during an accident without the need for human involvement.[38] Other passive safety systems include features such as natural circulation which would cool the reactor core without need for any external water replacement or coolant pumps.[39] Reactors such as the ACP100 designed by the China National Nuclear Corporation are also incorporating the Passive Decay Heat Removal System (PDHRS), Passive Emergency Core Cooling System (ECCS), Passive Containment Air Cooling System (PAS) and Reactor Automatic Depressurisation (RDP) system.[40]

In addition to the safety aspects, new SMR designs are also incorporating the concept of "security by design" to include enhanced security features that address concerns of sabotage, theft, attacks and proliferation. To minimise security risks during transportation of nuclear materials, some reactors are designed to operate for prolonged periods without refuelling.[41] These long refuelling intervals, by extension, minimise the handling and transportation of nuclear materials and improve nuclear security. Ideas of locating reactor units in more secure places such as underground or underwater are also being explored to protect them from natural calamities such as tsunamis.[42] This could also make the facilities difficult to break in, thus, minimising the chances of nuclear theft and sabotage.

38. S. Dotson, "The Promise of Small Modular Reactors, " Power-Eng.com, February 5, 2015. https://www.power-eng.com/nuclear/reactors/the-promise-of-small-modular-reactors-2/. Accessed on October 2, 2021.
39. n. 5.
40. "Advances in Small Modular Reactor Technology Developments: A Supplement to IAEA Advanced Reactors Information System (ARIS) 2020 Edition", Aris.iaea.org, https://aris.iaea.org/Publications/SMR_Book_2020.pdf. Accessed on October 1, 2021, p. 25.
41. "Benefits of Small Modular Reactors (SMRs)", energy.gov. https://www.energy.gov/ne/benefits-small-modular-reactors-smrs#:~:text=SMRs%20provide%20simplicity%20of%20design,as%20demand%20for%20energy%20increases. Accessed on October 2, 2021.
42. Ibid.

### *Flexibility of Installation and Siting*

SMRs would provide flexibility in options ranging from where they are sited to how many units are installed. With regard to siting, the portability of the reactors offers many advantages, including their placement in remote isolated areas that do not have grid infrastructure. In many such areas that are currently dependent on diesel generators, deployment of non-polluting SMRs would be of great value.

The advantage of easy dismantling of the reactor modules at the end of their lifetime, as well as addition of new modules when considered necessary, adds to the flexibility advantage.[43] Some SMRs are also being designed with a load-following design, which would let them downrate their electricity generation when the demand is low.

### *Baseload Power in Support of Renewables*

Nuclear power has a major advantage of being a baseload source of electricity. While there is a rapid deployment of renewable energy today, it does suffer from the disadvantage of intermittency. In this context, SMRs are seen as clean back-ups to renewable sources of energy to help stabilise the supply to the grid. So, SMRs could supplement solar, wind, small hydroelectric and tidal generation to ensure continuous supply of electricity. This combination would help overcome the challenge of storage that use of renewables still poses.

### *Ease of Technology Utilisation by Nuclear Newcomers*

Given that SMR designers claim a lesser requirement of operating staff due to the inherent safety systems, the reduced human resource requirements might make it easier for the technology to be absorbed without the need for elaborate human resource training. Would this then reduce the employment opportunities for nuclear scientists and engineers? The SMR proponents refute this to suggest that even as the number of people needed per reactor operations would reduce, construction of a larger number of SMRs would generate more

43. Ibid.

employment opportunities. A 2010 US study on the economic and employment impacts of SMR deployment revealed that a 100 MWe prototype SMR could help in creating about 7,000 jobs. It also estimated $1.3 billion in sales, and $404 million in earnings.[44]

### *Better Waste Management*

Different designs are experimenting with unique ways of dealing with nuclear waste. For example, the fast reactor designs have a higher fuel burn-up which leads to less nuclear waste generation; the travelling wave reactor design consumes the fuel that it breeds which helps in doing away with the need to remove spent fuel. Likewise, other SMR technologies have been trying to develop a thorium fuel cycle which too could reduce nuclear waste. Many reactor designs in this regard focus on developing concepts that could destroy isotopes in the spent nuclear fuel, thereby reducing their radioactivity.[45]

### *Better Resource Efficiency*

SMRs are also more resource efficient as they require less resources. With regard to land use, SMRs have a comparatively smaller physical footprint as they are less land intensive than traditional nuclear reactors. For example, a traditional nuclear power plant needs an emergency planning zone that extends up to 10 and a half miles around the plant. In comparison, SMRs require just about half a mile.[46] They are also time efficient as the deployment of reactors is estimated to take a shorter construction time. SMRs also require less maintenance. Their modular design can ensure that even if one of them goes offline for refuelling, the others can continue.[47]

44. Ibid.
45. "Small Modular Reactors", acee.princeton.edu. https://acee.princeton.edu/distillates/small-modular-reactors/. Accessed on October 12, 2021.
46. Ibid.
47. N. Cunningham, "Small Modular Reactors: A Possible Path Forward for Nuclear Power," *American Security Project*, October 2012. http://large.stanford.edu/courses/2015/ph240/tan2/docs/smr.pdf. Accessed on October 2, 2021.

## Potential Challenges of SMRs

### *Supply Chain Maturity: Long Way Off*

Modularity is being touted as the biggest advantage of SMRs. It is expected that SMRs will reap the benefits of serial factory manufacturing which would enable optimum standardisation of components. Preassembled modules will simplify on-site installation. The problem, however, remains that for factories to sink in substantive capital investment to create the infrastructure for such manufacturing, they would need the assurance of a sufficient number of orders. Only then will economic efficiencies emerge. Unless fleet orders are ensured, such production facilities are unlikely to come up. Such orders, however, appear a long way off in the future since it will take time before one particular design from the plethora currently being experimented with is sufficiently field-proven and accepted by a large number of companies, operators and regulatory bodies. And, it is only when this happens that mass production of the kind necessary to make SMRs economically viable can become a reality. Therefore, supply chain maturity of SMRs will take time, may be even a few decades, to emerge. It will depend on the kind of operating experience gained from the first series of such reactors. At this juncture, the economics of serial production of a design that has stabilised seems distant.

### *Shadow on the Price Advantage*

Claims of lower costs of SMRs can only accrue when economies of scale can be achieved. This, in turn, can only happen when the number of units being produced is so large that the price of an individual unit reduces with the spread of the cost of production over a large number of units. Large scale production of SMRs, however, requires a standardised reactor design, which appears a long way off, considering the number of designs being developed. In addition, mass production would require a large market with keen investors. However, in view of the uncertainty around SMRs, many investors are wary of investing in the technology. Andrew Storer, chief executive of the

Nuclear Advanced Manufacturing Research Centre, which predicts the market scenarios for nuclear power manufacturers, says that as for as supply chain companies are concerned, "We're advising people, 'Don't invest yet.'"[48] M.V Ramana, a professor of nuclear physics at Princeton University and a nuclear sceptic, too described the future of SMRs as "risky and expensive"[49] in a report he wrote on the Utah Associated Municipal Power Systems (UAMPS).He observed that the cost estimates for the UAMPS project increased from $3billion to over $6 billion in just five years.[50]

***Tussle Between Industry and Regulators***

Given the novelty of the concept, the SMR industry is, not surprisingly, facing new regulatory challenges ranging from licensing to liability. The current regulatory regimes are crafted primarily for traditional nuclear plants. In fact, some in the nuclear industry complain that the innovative approach of SMRs has not been adequately understood or appreciated by the regulators. Also, the multiplicity of designs has created a challenge for streamlining SMR licensing. It has led to diverse national regulatory approaches which will complicate deployment of the reactors globally. This challenge was well brought out by King Lee, the Harmony Programme director at the World Nuclear Association, when he said,

> Each country is responsible for regulating safety within their own jurisdiction, and [that] has resulted in different approaches…The UK and some countries in Europe tend to adopt a more goal-based safety regime, where the regulators set the goal and it's up to the licensee to demonstrate how that goal has been met…Some countries have a

48. n. 5.
49. Dave Levitan, "First US Small Nuclear Reactor Design Is Approved: Concerns About Costs and Safety Remain, However", *Scientific American*. September 9, 2020, https://www.scientificamerican.com/article/first-u-s-small-nuclear-reactor-design-is-approved/, Accessed on October 2, 2021.
50. Ibid.

> more prescriptive regulatory regime, like the US, where the regulator defines more precisely how to meet regulation. That is… part of different safety regulation philosophies between countries.[51]

Meanwhile, the regulators are exercising abundant caution. They argue that their task is agnostic to the size of the reactors and that it is to ensure the safety of the plant operations. Rather, they are being overcautious in view of the fact that many companies that have moved into the SMR space are new to the nuclear industry and not equipped to address all the concerns of the regulators. Also, since novel designs are being experimented with, they can brook no laxity in regulatory procedures. Kevin Lee, senior regulatory policy officer at the Canadian Nuclear Safety Commission stated this categorically when he said, "No nuclear regulator wants to take shortcuts on regulatory approval at the expense of safety, nor do SMR proponents…. "[52]

### *Problems with SMR Exports*

A big challenge for the export of nuclear reactors and their overseas deployment will remain getting the design approval of the regulators of the host country. It is the sovereign right of a nation to have its own regulatory oversight on nuclear imports. While in the case of traditional nuclear reactors, there is expected to be a national infrastructure that would be involved, in the case of SMRs that may be bought and installed, the proprietary design is likely to be sold as a black box. With design approval secured in one country not automatically becoming acceptable in others, since different regulators place emphasis on different issues, any demand in design changes would defeat the

51. Jason Deign, "Regulatory Harmonization: An Upcoming Hurdle for SMRs?" greentechmedia.com, February 15, 2021. https://www.greentechmedia.com/articles/read/regulatory-harmonization-an-upcoming-hurdle-for-smrs. Accessed on October 4, 2021.
52. Jason Deign, "Regulatory Harmonization: An Upcoming Hurdle for SMRs?" greentechmedia.com. February 15, 2021. https://www.greentechmedia.com/articles/read/regulatory-harmonization-an-upcoming-hurdle-for-smrs. Accessed on October 4, 2021.

advantage of modularity and stability of supply chains with pre-fab reactors that SMRs seek to exploit. It would also drive up the cost of reactors, thereby, defeating the purpose of the SMRs.

Harmonisation in licensing requirements and processes could facilitate their deployment in different countries without significant adaptations to meet national regulations. Design changes may then only be necessary to meet site specific characteristics. The procurement of different components in local markets would also be possible, which will, in turn, enable the creation of global markets and global supply chains. As such, harmonisation will play a central role in supporting the economies of series, indispensable for the competitiveness and commercial viability of this technology. But, it remains to be seen whether international regulatory bodies, governments and industries will be able to coordinate their efforts to develop international standardisation of licensing and regulation. Some such efforts have been undertaken by programmes such as the Cooperation in Reactor Design Evaluation and Licensing (CORDEL).[53] The progress, however, has been slow.

### *Liability Issues*

Liability in case of an accident at a nuclear plant is a matter of great concern and contention. These are further exacerbated in the case of SMRs, especially given the many new dimensions of technology, designs material and fuel. Some have even raised doubts on whether existing liability conventions can apply to factory production of units as in the case of SMRs.

The maximum concerns in this regard have been raised in the context of FNPPs. The first of these pertains to the very definition of such reactors under the existing conventions. For instance, the Convention on Nuclear Safety (CNS) defines a "nuclear installation" as "any land-

53. "Facilitating International Licensing of Small Modular Reactors", Cooperation in Reactor Design Evaluation and Licensing (CORDEL) Working Group of the World Nuclear Association, August 2015. http://www.world-nuclear.org/uploadedFiles/org/WNA/Publications/Working_Group_Reports/REPORT_Facilitating_Intl_Licensing_of_SMRs.pdf. Accessed on October 3, 2021.

based civil nuclear power plant under its jurisdiction". Can this then apply to FNPPs? Experts differ on whether a floating nuclear power plant at shore or offshore may be considered a nuclear installation under the international third-party nuclear liability conventions.

Questions have also been raised on what happens when the FNPP navigates through different maritime zones and high seas. The CNS only refers to the carriage of nuclear substances, i.e. nuclear fuel and radioactive products and waste. A discussion on the liability regime applicable during the carriage is necessary. This is especially necessary to facilitate the insurance coverage of such installations and protect potential victims in case a nuclear incident occurs during the journey.

### *Safety, Security and Safeguards (3S) Concerns*

Given their new physical layouts, designs, use of novel fuels and increased digitisation, it is expected that SMRs would require new 3S protocols.

From the safety perspective, technical studies have found benefits in the new features of inherent and passive safety systems. However, on security, concerns have been raised regarding the possibility of malafide activities during transportation. Furthermore, as SMRs are expected to be deployed in remote locations, safeguards efforts will need to expand. It is also argued that for sealed core SMRs, general safeguards procedures may not suffice.[54]

Such challenges put new burdens on the numbers and required qualifications of IAEA inspectors. In an attempt to mitigate some of the 3S concerns, the IAEA has been encouraging countries to work with them from "an early stage in reactor design, ... safeguards considerations can be embedded into the design of these reactors, so that nuclear verification can be performed in the most effective and efficient way with minimal

---

54. G. Bentoumi, A. Chaudhuri, et al., "Safety And Security For Small Modular Nuclear Reactors In Canada", Conferences. Iaea. https://Conferences.Iaea.Org/Event/181/Contributions/15437/Attachments/9190/12388/Paper_-_Safety_Security_For_Smr_In_Canada_-_G._Bentoumi_Final.Pdf. Accessed on October 28, 2021.

burden on the operator…".[55] To support countries in doing so, the IAEA has introduced the "Safeguards by Design" (SBD) guidance series[56] which can help ensure that consideration of international safeguards is included in the design process along with other choices of economic, operational, safety and security factors. However, this approach is challenged by the IAEA's lack of direct communication with the nuclear facility designers to sensitise them to the 3S risk complexity in nuclear reactors.

## India: The Way Ahead with SMRs

Of the 22 nuclear reactors operational in India today, the majority of the indigenously built ones have a capacity rating of 220 MWe. India also has the experience of having built an 85 MWe reactor for its nuclear submarine. This shows that India has the capability to design, build and operate small reactors. While it has not utilised a modular factory process so far, the industry involved in manufacturing nuclear equipment can be expected to make this possible in case the nuclear establishment is keen to build such reactors.

Cognisant of the potential of SMRs, India's DAE has design teams working on the technology.[57] Speaking of the SMR prospects for the country, K N Vyas, secretary, DAE and chairman, Atomic Energy Commission, had stated in 2019,

> Carrying out the design of new reactor systems and refinement in the already performed design is an ongoing process, which is always under focus to improve the designer's capability. SMRs also need

55. Carmen Cabañas, "Safeguarding the Nuclear Future: Small Modular Reactors", IAEA.org, September 23, 2021. https://www.iaea.org/newscenter/news/safeguarding-the-nuclear-future-small-modular-reactors. Accessed on October 5, 2021.
56. "Safeguards by Design Guidance", IAEA.org. https://www.iaea.org/topics/assistance-for-states/safeguards-by-design-guidance., Accessed on October 6, 2021.
57. "DAE working on Small Modular Reactors: KN Vyas", *Nuclear Asia*, November 21, 2019, https://www.nuclearasia.com/news/dae-working-small-modular-reactors-kn-vyas/3307/. Accessed on October 16, 2021.

some technology development to fill-up gap areas. The process of technology development also needs to be completed before the task related to SMRs can be taken up in a more serious manner.[58]

While the DAE is keeping abreast of SMR developments, its current priority is technological advances on the current stream of reactors. Having graduated from 220 MWe to 540 MWe, India has developed indigenous technology for 700 MWe reactors in an attempt to reach higher capacity reactors, and is now building a string of such reactors across various sites. In order to ensure a rapid expansion of nuclear electricity generation, the country has been inclined towards importing nuclear reactors of larger capacities.

Can SMRs then be attractive for a country like India which has a grid size that is capable of absorbing large reactors and where the demand for electricity is only expected to grow? Indeed, while the rationale for large reactors is stronger in India, SMRs too could be useful in a few scenarios. One of these could be to replace old coal plants that need to be decommissioned. In this regard, Srikumar Banerjee, former chairman, Atomic Energy Commission, had rightly stated, "Some of the retiring thermal power plants can be replaced with small modular reactors."[59] This would also help in combatting air pollution problems plaguing the country.

A second utility for SMRs may be found in powering remote areas which lack the necessary electricity infrastructure. Given that a 2018 survey by the central Rural Development Ministry had revealed that over 14,700 villages in India did not have electricity for domestic use[60],

58. Ibid.
59. "Small Modular Nuclear Reactors are Now Beautiful for Plant Makers". Energyworld.com, February 25, 2021. https://energy.economictimes.indiatimes.com/news/power/small-modular-nuclear-reactors-are-now-beautiful-for-plant-makers/81209735. Accessed on August 16, 2021.
60. Bhasker Thripati, "5 Million Indian Households Have Meters But No Electricity", *BloombergQuint*, November 1, 2018. https://www.bloombergquint.com/global-economics/15-million-indian-households-have-meters-but-no-electricity. Accessed on October 16, 2021.

SMRs could help provide electricity to some of these villages in remote areas. Thirdly, given the high induction of renewable power in a thrust to move towards green technologies, SMRs could supplement this by being a dependable base load and clean source of power.

Lastly, as explained by Mr. Sunil Ganju, head, Institutional Collaboration and Progress Division (ICPD) at the Department of Atomic Energy (DAE), "…SMRs of molten salt categories can use our vast thorium resources."[61] It may be recalled that India's three-stage power programme has plans to move towards large-scale utilisation of domestic thorium reserves through the transition from PHWRs to FBRs to thorium reactors. With many global SMR designers working on molten salt technologies, this may provide an opportunity for India to fast-track its plans for thorium exploitation.

In view of the above, there are potential uses for SMRs in India. But, given the country's electricity requirements, these need not feature at the top in the DAE's priority list. Developed countries that are engaged with SMRs are doing so at a stage where their electricity demand and population growth are mostly stagnant. The new technology development, then, is actually a way for the nuclear industry to keep itself gainfully occupied and with a view to popularise a new nuclear technology for countries where reactor construction has stagnated or where there is potential to export them to nuclear newcomers. For India to divert any large part of its nuclear resources towards such a technology would be a diversion from its focus, which should be on quick construction of planned nuclear reactors in order to enhance the country's electricity supply with an environmentally friendly source. Long-term policy support is needed to bring the planned reactors of larger sizes to fruition and to do so within specified time schedules.

Nevertheless, the country must remain engaged with SMR R&D since these can find a role in the many needs of an economically growing country. For India, SMRs may be particularly useful for meeting the

61. "Factors that Augur Well for Small Modular Reactors", *Nuclear Asia,* March 2, 2021. https://www.nuclearasia.com/features/factors-that-augur-well-for-small-modular-reactors/4115/. Accessed on October 28, 2021.

power requirements of difficult areas such as Ladakh or Siachen where transporting energy is a huge challenge. It can also be imagined that such reactors will be needed in the future for space propulsion, mining of asteroids or other celestial bodies or even for establishing settlements on other planets.

Unfortunately, India currently does not have a private nuclear industry that could work on this technology independently. Perhaps, the government may consider incentivising private players to move on this front since it does plan to involve private industry with the Nuclear Power Corporation of India Limited (NPCIL) for reactor construction and operations. Encouraging them to have joint ventures with foreign companies engaged in SMR R&D may hasten their induction. It may be recalled that in November 2019, Rosatom of Russia had expressed interest in teaming with India to produce 50-100 MW reactors for exports.[62] This could be a worthwhile venture and will help establish India in the field of nuclear exports. Meanwhile, the DAE-NPCIL resources at this stage of the nuclear power programme can continue to focus on the rapid completion of its planned 700 MWe reactors. Construction of these reactors in fleet mode would itself offer the advantages of modular construction and be more in tune with the country's immediate requirements.

62. Dipanjan Roy Chaudhury, "Rosatom Expresses Interest in Teaming up with India to Produce Small and Medium Sized Reactors", *Economic Times*, November 8, 2019.

# 4. Military Applications of Nuclear Power Plants

*Anil Chopra*

The first use of nuclear energy was in the form of the atomic bomb. Dropped over Hiroshima and Nasgasaki, it demonstrated the destructive potential of nuclear fission. In the decade thereafter, the focus was changed towards harnessing the atom for peaceful uses. Nuclear energy then became the buzzword.

Besides reactors for electricity generation for civilian grids, nuclear reactors have also been found useful for military platforms, especially on sea-going vessels, giving them greater power for a longer time. In fact, nuclear power plants for maritime applications began evolving in the United States in the 1940s. The United States and Soviet Union have had nuclear powered submarines since the early 1950s.[1] Unlike the diesel-electric powered submarines which have to surface frequently to breathe air to run the generators for charging the batteries, submarines with nuclear propulsion are independent of such requirement. The much more efficient power generation also allows higher speed and long periods of operation without having to refuel. In fact, they may not be refuelled for their entire operational life of typically around 25 years. Even the most advanced electric batteries used in the relatively newer diesel submarines can at best remain submerged for just a few days at slow speed. This figure could reduce to just a few hours at high

1. World Nuclear Association, "Nuclear-Powered Ships", Updated September 2021. https://world-nuclear.org/information-library/non-power-nuclear-applications/transport/nuclear-powered-ships.aspx. Accessed on October 17, 2021.

speeds.[2] The design of marine-type reactors is quite different from the electricity generation reactors used for the national power grid supply.

This paper examines the military applications of nuclear power plants. It examines their historical dimension, their present status and likely future applications.

## Small Nuclear Power Reactors

The International Atomic Energy Agency (IAEA) defines 'small' as under 300 Megawatts electric (MWe), and up to about 700 MWe as 'medium'.[3] There is interest in small nuclear power reactors for both military and civil applications. For long, the navy has been using nuclear powered aircraft carriers and submarines. There is also interest for both military and civilian uses for portable small power plants for use in remote locations. While small reactors require special technologies, they are less cost intensive, especially for transmission. Experience from nuclear powered submarines is available already. Some of these have been generating less than 190 MW power. Greater demand would also mean economy of scale. Modern small reactors are being designed to be modular. There are also very small units which are about 15 MWe, especially for remote communities.

## Military Small Power Reactors

For long, the US Army has been using mobile and static small reactors. These were meant to power remote air/missile defence radar stations in Alaska, Greenland, and Antarctica, for providing electricity and heating. A merchant ship called *Liberty* was converted into the first experimental Floating Nuclear Power Plant (FNPP) in mid-1960s. The

2. Amit Kumar Mukherjee, "A Birds Eye View of Submarine Propulsion—Past, Present and Future", Mission Victory India, August 6, 2021. https://missionvictoryindia.com/a-birds-eye-view-of-submarine-propulsion-past-present-future/. Accessed on October 17, 2021.
3. World Nuclear Association, "Small Nuclear Power Reactors", Updated September 2021. https://www.world-nuclear.org/information-library/nuclear-fuel-cycle/nuclear-power-reactors/small-nuclear-power-reactors.aspx. Accessed on October 17, 2021.

truck-mobile, and air-transportable plant was of the size of a standard shipping container. It was operated in the Panama Canal Zone from 1968-77.[4]

The US armed forces are looking at reducing power supply vulnerabilities and operating costs, and also reducing petroleum dependence. Mini Nuclear Power Plants (MNPPs) would be portable by truck or large C-17 class aircraft and operate unrefuelled for 10-20 years. More recently, they have been looking at mobile Small Modular Reactors (SMRs) for rapid response scenarios. These will be relevant in the case of an electrical grid attack or for augmenting power requirements. These would also be handy during Humanitarian Assistance and Disaster Relief (HADR) operations. The reactors should need minimal operating crew. The smaller ones, weighing less than 40 tons, could generate around 10 MWe for around three years without refuelling. Safety and reliability will be crucial. Fairly tight installation and shutdown timelines have been specified by the Department of Defence (DoD). The environmental impact has been factored in. In March 2021, the DoD announced that "Project Pele[5]" was on track, and full-power testing of the mobile reactor was expected in 2023. Outdoor mobile testing of the prototype micro-reactor is likely in 2024. US firms, such as NuScale, are also working on a floating version of an SMR. These could be used for the power requirements of coastal areas when anchored close to the shore. There are other SMR projects by Canadian and Japanese firms. Rolls-Royce has been developing naval reactors since the 1950s, and is also working on small test reactors.

Even the Soviet Union built the Pamir-630D truck-mounted small air-cooled 0.6 MWe nuclear reactors. These were discontinued later. Russia now has small transportable 2.5 MWe and 6 Megawatts hour (MWh) nuclear reactors. Russia is also developing small mobile nuclear

4. "MH-1A Nuclear Reactor at Fort Belvoir". http://www.virginiaplaces.org/energy/nuclearbelvoirmh1a.html. Accessed on October 19, 2021.
5. Jeff Waksman, Programme Manager, Project Pele Overview, Department of Defence, February 2021. https://ric.nrc.gov/docs/abstracts/waksmanj-w15-hv.pdf Accessed on October 7, 2021.

power plants for the military in the Arctic. These can be moved in IL-76 aircraft and Mi-26 helicopters. The USA, Russia and China also have Light Water Reactors (LWRs) quite like the water cooled naval reactors. The plant can be mounted on a large barge. Some of the Russian icebreakers also use such plants.

## Nuclear Reactors for Deployed Troops

During their operations in Iraq and Afghanistan, the US armed forces had difficulty in getting huge quantities of fuel for the power needs of remote operating bases. More than 900,000 gallons of fuel a year were used for power generation.[6] It was significant, though relatively small compared to 6.7 million gallons used by aircraft and ground vehicles. Seventy percent of the logistics effort for the troops in the wars in Afghanistan and Iraq was related to energy and water supply.[7] Given this experience, it is not surprising that the US military has sought quick deployable SMRs. The requirement was to have reactors that could be transported in C-17 aircraft and then moved on a flat-top trailer. A brigade size requirement would be for 10 MW of electric power. The small reactors in field conditions would also power hydrogen electrolysis units to generate hydrogen for vehicles. This would permit conversion of more ground vehicles to run on hydrogen, thus, reducing the requirement of other fuels.

Super small reactors with radioisotope thermoelectric generators are being considered as the answer. The National Aeronautical and Space Administration (NASA) reportedly uses some of these to power satellites and other spacecraft. In 2020, the Pentagon awarded three contracts for mobile, small nuclear reactors, to provide nuclear power for American

---

6. Patrick Tucker, "US Military Eyes Tiny Nuclear Reactors for Deployed Troops", *Defence One*, January 24, 2019. https://www.defenseone.com/technology/2019/01/us-military-eyes-tiny-nuclear-reactors-deployed-troops/154406/. Accessed on October 28, 2021.
7. Richard B. Andres and Hanna L. Breetz, "Small Nuclear Reactors for Military Installations: Capabilities, Costs, and Technological Implications", Strategic Forum, National Defence University, February 2011, p. 4. https://ndupress.ndu.edu/Portals/68/Documents/stratforum/SF-262.pdf. Accessed on October 20, 2021.

forces at home and abroad.[8] The 2-10 MWe range reactors are expected to be available by 2023. They could be deployable by 2027.[9] The military would also require these for homeland use so as to avoid using the civil power grids that could be susceptible to cyber attacks. A technical study has indicated that the annual energy requirements of a majority of US military installations could be met by nuclear power capacity of 40 MWe or less.[10] Alternatively, smaller units of 2 to 10 MWe could act as critical systems power back-up in the case of grid failure. Many American companies[11] such as NuScale, Westinghouse, and Ultra Safe Nuclear are all developing under 10 MWe reactors. Sweden's LeadCold and a UK consortium led by URENCO are other players in the segment.

The technical challenges of keeping the radioactive fuel safe and operational in battlefield conditions would have to be addressed. The nuclear reactors and/or fissile material will also have to be safeguarded from possible theft or capture by terrorist groups. Though the nuclear material used is unlikely to be usable for a nuclear weapon, some kind of dirty bomb could be crafted. Security during the move of the reactor would have to be very high. For refuelling the reactor after 3 or 10 years, it would have to be moved back to a secure place.

SMRs are likely to face substantial "not-in-my-backyard" battles. Forward located reactors could be important targets for the adversary's mortar or air attack. Since forward bases could also be subject to capture, there would be a need for contingency plans in case the reactors fall into enemy hands.

Many analysts feel that the small reactors would play an important role in the second nuclear era. However, even if they prove their

---

8. Aaron Mehta, "Pentagon Awards Contracts to Design Mobile Nuclear Reactor", *Defence News*, March 9, 2020. https://www.defensenews.com/smr/nuclear-arsenal/2020/03/09/pentagon-to-award-mobile-nuclear-reactor-contracts-this-week/. Accessed on October 20, 2021.
9. Aaron Mehta, "Portable Nuclear Reactor Project Moves Forward at Pentagon", *Defence News*, March 23, 2021. https://www.defensenews.com/smr/energy-and-environment/2021/03/23/portable-nuclear-reactor-project-moves-forward-at-pentagon/. Accessed on October 20, 2021.
10. Mehta, n. 8.
11. Ibid.

commercial viability for civilian purposes, militaries will have to factor in many operational security issues besides the problems of reliability, performance, and proliferation resistance.

## Nuclear-Powered Submarines

Nuclear-powered submarines have small reactors that provide propulsion. Their advantage over conventional diesel-electric submarines is based on the fact that they are independent of air and do not need to surface off and on to run the diesel generator to charge the batteries. Also, the nuclear reactor generates significantly high power and, can operate at high speed for long durations, thus, giving it a nearly unlimited range of operations, albeit restricted only by replenishments of other consumables like food, etc. Modern nuclear submarines do not need refuelling in their entire lifespan of nearly 25 years[12], but there may be a need for routine maintenance of other systems. On the other hand, as stated earlier a diesel-electric submarine can be submerged for only about 2-3 days operating at slow speed, and just a few hours when operating at high speed. However, nuclear submarines cost much more and only a few countries have the technology and means to build and sustain them. There are also nuclear safety issues.

**Fig 1: A Typical Nuclear Submarine Layout**

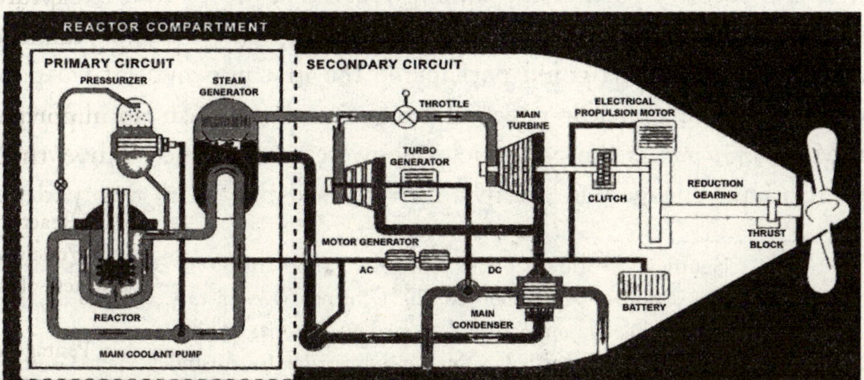


12. Ensigns Michael Walker and Austin Krusz, "There's a Case for Diesels", US Naval Institute, June 2018. https://www.usni.org/magazines/proceedings/2018/june/theres-case-diesels. Accessed on October 29, 2021.

The idea of nuclear submarines first evolved during World War II. The United States and the Royal Navy were the first to develop a nuclear propulsion plant. The US Congress authorised the construction of the first nuclear submarine in 1951. The Westinghouse Corporation built the reactor. Named the USS *Nautilus*, it was commissioned in September 1954.[13] The British and Soviets followed. The Soviet Navy got its first nuclear submarine in 1958. The first British nuclear-powered submarine had an American reactor. The subsequent Valiant class had Britain's own Rolls-Royce's PWR1 nuclear plant. Nuclear submarines became even more important with nuclear missiles and the need for longer submerged loiter time, thus, remaining difficult to detect. The USS *George Washington* with 16 Polaris A-1 missiles was the first US nuclear-powered and nuclear ballistic missile armed submarine (SSBN) to be commissioned in late 1960. The Soviets also commissioned their first SSBN around the same time. However, their American equivalent entered service only in 1967, by when the USA already had 41 SSBNs. The Soviet Union, later Russia, finally built more SSBNs than all other nations put together.

The six countries that currently operate SSBNs are the United States, Russia, China, France, the United Kingdom, and India. A few other countries are also building nuclear-powered submarines. The new trilateral security partnership among Australia, the United Kingdom and the United States (AUKUS), emerged in September 2021.[14] As part of the deal, Australia could perhaps be the first non-nuclear weapon state to operate a nuclear-powered submarine. The British Royal Navy's HMS *Conqueror*, is the only nuclear-powered submarine in the world ever to have engaged an enemy ship.[15] It used two Mark 8 torpedoes

13. "USS Nautilus—World's First Nuclear Submarine—is Commissioned," History.com Editors, September 28, 2021. https://www.history.com/this-day-in-history/uss-nautilus-commissioned. Accessed on October 21, 2021.
14. Julia Masterson, "U.S., UK Pledge Nuclear Submarines for Australia", Arms Control Association, October 2021. https://www.armscontrol.org/act/2021-10/news/us-uk-pledge-nuclear-submarines-australia. Accessed on October 17, 2021.
15. Blake Stilwell, "This British Sub Hoisted its own Jolly Roger after Sinking an Argentine Cruiser", *We Are The Mighty*, August 14, 2021. https://www.

during the 1982 Falklands War to sink the Argentinian cruiser ARA *General Belgrano*.

Nuclear reactors used in submarines usually use highly enriched uranium as fuel (often greater than 20 per cent). The reactor is positioned inside the submarine's pressurised hull. The reactor also supports other on-board systems such as the desalination plant, air conditioning and purification. Diesel generators remain for back-up. The nuclear submarine generates a large quantity of heat. The high thermal wake that moves up to the sea surface can be detected by infrared sensors. There is, thus, a need to cool the reactor. The continuously running reactor also creates steam noise that can be detected by sonars. Both these issues reduce the stealth characteristics of the submarine. Additional cooling and noise suppression systems are, hence, incorporated. In any case, the huge advantage of higher speed makes it more difficult to target a nuclear-powered submarine. The Soviet Project 661 submarines are known to have a top submerged speed of as high as 70 km per hour.[16]

## Operational Nuclear Submarines

Russia and the USA have had a large number of nuclear-powered submarines over the years. Both continue to have the largest number of operational nuclear submarines. As of 2020, the USA had a total of 53 fast attack submarines, 32 of which are the Los Angeles class submarines, which are more than any other class in the world. These have a top speed of 69 km per hour, and an operating depth of 650 ft (200 m). The United States Navy (USN) has 14 Ohio class SSBNs and four cruise missile submarines (SSGNs). It also has three, $3 billion a piece, Seawolf class fast-attack submarines (SSNs). The USN's

wearethemighty.com/mighty-trending/this-british-sub-hoisted-its-own-jolly-roger-after-sinking-an-argentine-cruiser/. Accessed on October 21, 2021.

16. Joseph Trevithick, "The Soviet's 'Golden Fish' Missile Submarine Still Holds The Record As The World's Fastest", *The Drive*, February 13, 2019. https://www.thedrive.com/the-war-zone/26475/the-soviets-golden-fish-missile-submarine-still-holds-the-record-as-the-worlds-fastest. Accessed on October 17, 2021.

latest submarines are of the Virginia class (SSN-774). These bring on board the latest stealth, intelligence gathering, and weapon systems technologies. They have "sail-by-wire" capability that greatly reduces submarine acoustic and heat signatures and allows quieter operations even in shallow waters, thus, obviating the need to surface.[17] They are replacing the older Los Angeles class submarines. The USN plans to induct the Virginia class submarines till the mid-2040s, and, therefore, they would be in service till the 2070s.[18]

The Russian Navy has an estimated 58 submarines that include 11 SSBNs, 17 SSNs, 9 SSGNs, and 21 diesel-electric attack submarines (SSKs). The Russian Navy plans to replace both its SSNs and SSGNs with the multi-role Project 885-M Yasen class submarines.[19] These have recently begun entering service, and are comparable to the US Navy's Block-V Virginia class submarines. The Russian government remains committed to the modernisation programme to maintain its nuclear deterrent. The new Borei-K class nuclear submarines will be capable of firing long-range cruise missiles. A new long-range nuclear-powered torpedo, the Status-6 or "Poseidon" is fitted with a 100-megaton nuclear warhead designed to create radioactive tsunamis. In April 2019, Russia launched the Project 9852 Belgorod, a submarine that can launch the Status-6 torpedo.

Just two of the seven built Trafalgar class SSNs of the Royal Navy are still in service in the UK. They are being replaced by the Astute

17. Kris Osborn, Warrior Haven, Submarine Surge: Why the Navy Plans 32 New Attack Subs by 2034", Fox News, March 25, 2019. https://www.foxnews.com/tech/submarine-surge-why-the-navy-plans-32-new-attack-subs-by-2034. Accessed on October 17, 2021.
18. Loren Thompson, "Five Reasons Virginia-Class Subs are the Face of Future Warfare", *Forbes*, May 6, 2014. https://www.forbes.com/sites/lorenthompson/2014/05/06/five-reasons-virginia-class-subs-are-the-face-of-future-warfare/?sh=26244fa413fe. Accessed on October 17, 2021.
19. Peter Suciu, "How the US's and Russia's Newest Attack Submarines Stack up", *Insider*, August 9, 2021. https://www.businessinsider.com/how-russian-yasen-class-us-virginia-class-attack-submarines-compare-2021-8?IR=T. Accessed on October 17, 2021.

class[20], of which, four out of the seven planned are already active. The Royal Navy also has four Vanguard class SSBNs in service. These will one day be replaced by the Dreadnought class: two out of the planned four are under construction and will enter service in 2028. They will have a service life of around 35 to 40 years.

Four French SSBNs of the newer Le Triomphant class are in the service of the French Navy. The first of the planned six Barracuda class nuclear attack submarines was commissioned on November 6, 2020. These incorporate many improvements, and extend the time between refuelling and complex overhauls from 7 to 10 years, allowing higher operational availability. The French are also developing the SNLE 3G[21], third generation nuclear ballistic missile submarine as a replacement for the Triomphant class, beginning around 2035, and to be in service till 2090.

The People's Liberation Army Navy (PLAN) operates a fleet of 68 submarines. China's submarine fleet is expected to grow to 76 by 2030.[22] China was the pioneer in Asia and the fifth in the world to build a nuclear-powered submarine. The PLAN has three Type 091 nuclear submarines and six Type 093 nuclear-powered attack submarines in service. The Type 091 is extremely noisy, has poor radiation shielding, and cannot launch missiles while submerged. These are being replaced gradually. The Type 095, Tang class SSN or Type 096 (SSBN) will be the third generation nuclear-powered attack submarines.[23] These will have reduced acoustic

20. "SSN Astute Class Nuclear Submarine", *Naval Technology*, October 2021. https://www.naval-technology.com/projects/astute/. Accessed on October 25, 2021.
21. Xavier Vavasseur, "France Launches Third Generation SSBN Program—SNLE 3G", *Naval News*, February 21, 2021. https://www.navalnews.com/naval-news/2021/02/france-launches-third-generation-ssbn-program-snle-3g/. Accessed on October 28, 2021.
22. H I Sutton, "Chinese Navy Steps Closer to New Generation of Nuclear Submarines", *Forbes*, June 19, 2020. https://www.forbes.com/sites/hisutton/2020/06/19/chinese-navy-gets-closer-to-new-generation-of-nuclear-submarines/?sh=38250ddc229e. Accessed on October 17, 2021.
23. H I Sutton, "First Image of China's New Nuclear Submarine Under Construction", *Naval News*, February 1, 2021. https://www.navalnews.com/naval-news/2021/02/first-image-of-chinas-new-nuclear-submarine-under-construction/. Accessed on October 17, 2021.

signature, a more advanced nuclear reactor, and better sensors. They may be employed as undersea escorts for future PLAN aircraft-carrier task forces.

Meanwhile, Brazil is also trying to develop its first 6,000-ton attack submarine. The Royal Australian Navy would get to acquire nuclear-powered submarines under the AUKUS agreement.

The Soviet K-43, Charlie class nuclear-powered cruise missile submarine, which had been commissioned in November 1967, was later leased to the Indian Navy and was operated as the INS *Chakra* from 1988 to 1991. It was manned by both Indian and Soviet crews, but the Indians were not allowed into the missile and reactor compartments. The lease was, therefore, not extended, but it did give the Indians experience then and helped them to build the Arihant class nuclear submarines. Meanwhile, the INS *Chakra II*, a Russian Akula class attack submarine, was leased by India for 10 years starting December 2011. India then began building its own Arihant class nuclear-powered ballistic missile submarines. The INS *Arihant* was commissioned in August 2016.[24] It was the first ballistic missile submarine built by a country other than the P-5. The 6,000-ton INS *Arighat*, the second in the series, is already on sea trials and is expected to be commissioned in August, 2022. It can carry 12-15 ballistic missiles. The next two are planned to be commissioned in 2023 and 2025 and would be of the 7,000-ton class. The Indian Government has approved six new nuclear-powered attack submarines (SSNs) under Project 75 Alpha. The current timeline for construction to commence is 2023-24, with the delivery of the first submarine around 2032. In 2019, India contracted yet another Russian Akula class nuclear-powered attack submarine christened the *Chakra III*.[25] It is expected to

24. Bhaswar Kumar, "INS Arihant Commissioned in Aug, but not 'Fully Ready' for Patrols: Report", *Business Standard*, October 18, 2016. https://www.business-standard.com/article/current-affairs/ins-arihant-commissioned-in-aug-but-not-fully-ready-for-patrols-report-116101800198_1.html. Accessed on October 26, 2021.
25. PTI, "INS Chakra, India's Lone Nuclear-Powered Attack Submarine, Returns to Russia", *The Mint*, June 4, 2021. https://www.livemint.com/news/india/ins-chakra-india-s-lone-nuclear-powered-attack-submarine-returns-to-russia-11622825015291.html. Accessed on October 17, 2021,.

be operational with the Indian Navy by 2025, and operate for around 10 years. The much heavier 13,500-ton class S5 submarines are planned to be armed with up to 12 or 16 ballistic missiles and have multiple independently targetable reentry vehicles. The construction is meant to begin in 2025.

Some serious nuclear and radiation accidents have taken place on nuclear submarines. There have also been cases of reactor core damage during maintenance work or refuelling in shipyards.

Dismantling decommissioned nuclear-powered submarines has its complexities. All major submarine operators have had to go through decommissioning. Step one is to defuel. The reactor section is then detached from the vessel. This section may be disposed of in a shallow land burial.

### Floating Nuclear Power Plants

Any marine nuclear propulsion plant has to be designed for greater self-sustenance and longer maintenance cycles so that it can increase sea-time far from its home port. Marine nuclear reactors have been used in the past to provide power supply at remote coastal areas including in the Arctic and Antarctica. These then function as Floating Nuclear Power Plants (FNPPs). One such plant has been used in the Panama Canal Zone to provide power to the Canal Zone. Russia has also built an FNPP at St Petersburg for its northern and far eastern territories. These have evolved from the nuclear plants on some of their ice breakers. China also has two FNPP projects.[26] Construction of the first started in November 2016. These are meant to power offshore oil and gas exploration and production units. South Korea and Canada are other countries that are developing FNPPs.

The excellent nuclear safety record of nuclear powered ships, combined with the need to reduce greenhouse gas emissions has led to renewed interest in marine nuclear propulsion. Marine nuclear plants would have

26. Trym Aleksander Eiterjord, "Checking in on China's Nuclear Icebreaker", *The Diplomat*, September 5, 2019. https://thediplomat.com/2019/09/checking-in-on-chinas-nuclear-icebreaker/. Accessed on October 28, 2021.

a great role in producing hydrogen or ammonia to power ships in the coming times. As of now, Russia's 61,900-ton *Sevmorput* is the only nuclear-powered freighter in service.[27] More such ships could be plying the trade routes soon. Some suggest that mid-sized 100 MW modular molten salt reactors could be very suitable for marine propulsion. Large bulk carriers on a few long routes could be the beginning. There could also be nuclear tugs to take conventional ships across oceans. However, many of these ideas are yet to be tested for their feasibility and a cost assessment will need to be made of how the risks of safety and security would be handled.

### Nuclear Aircraft Carriers

Only the United States and France have ever built nuclear aircraft carriers till now. France commissioned its only aircraft carrier, the 42,000-ton *Charles de Gaulle* in 2001. A second one is planned. The USN has had many such carriers. The USS *Enterprise* that operated from 1962 to 2012 was powered by eight reactor units, and till date remains the only aircraft carrier to have more than two nuclear reactors. Currently, the USN operates 11 carriers, and all are nuclear-powered. The biggest and the latest are of the 110,000-ton, Gerald R. Ford class, the first of which joined service in 2017. Nine more are planned. The United Kingdom rejected nuclear powered aircraft carriers very early due to the high cost. Construction of the already behind schedule China's fourth aircraft carrier has recommenced. This vessel is likely to be China's first nuclear-powered one. It is likely to be a 100,000-ton carrier. It would be China's first flat flight deck carrier with catapults. This will add to the country's nearly one dozen nuclear-powered submarines.

The Soviet Union did not build nuclear-powered aircraft carriers but has heavy nuclear-powered guided missile cruisers. These are the

---

27. "Nuclear-Powered Ships", World Nuclear Association, September 2021. https://world-nuclear.org/information-library/non-power-nuclear-applications/transport/nuclear-powered-ships.aspx. Accessed on October 28, 2021.

Kirov class, nuclear-powered guided missile cruisers[28] which are among the largest and heaviest warships operating in the world.

The USN also built similar cruisers, but these were retired before the year 2000. In 1961, the USN's USS *Long Beach* became the world's first nuclear-powered surface combatant ship. The last USN nuclear-powered cruisers were commissioned in 1980, and all were retired by 1999, as they were not found to be operationally cost-effective.

The high cost of nuclear technology and maintenance means that very few military powers can afford nuclear submarines or ships. Meanwhile, in 2020, the Pentagon[29] issued contracts to start the design work on small mobile nuclear reactors for power generation for the American military's use at homeland bases and on operational deployments abroad.

## Other Nuclear Military Ships

The Soviet *SSV-33 Ural* was a command and control naval ship with nuclear propulsion. It was used for electronic intelligence, and missile and space tracking, but was discontinued due to the heavy operating costs. The *Poseidon*[30] is a nuclear-powered and nuclear-armed unmanned underwater vehicle under development in Russia, with the capability to deliver both conventional and nuclear warheads.

Russia has nuclear-powered and nuclear-armed unmanned underwater vehicles. In the 1960s, the USA also embarked upon building a few experimental nuclear-powered civil merchant ships.[31] These were

28. Benjamin Brimelow, "Why the Soviet Union's Nuclear-Powered Cruisers Spooked the US into Bringing Back its Battleships", *Business Insider*, March 15, 2021. https://www.businessinsider.in/international/news/why-the-soviet-unions-nuclear-powered-cruisers-spooked-the-us-into-bringing-back-its-battleships/articleshow/81514327.cms. Accessed on October 27, 2021.
29. Mehta, n. 8.
30. Franz-Stefan Gady, "US Intelligence: Russia's Nuclear-Capable 'Poseidon' Underwater Drone Ready for Service by 2027", *The Diplomat*, March 26, 2019. https://thediplomat.com/2019/03/us-intelligence-russias-nuclear-capable-poseidon-underwater-drone-ready-for-service-by-2027/. Accessed on October 27, 2021.
31. Dan Wang, "When America Dreamed of a Nuclear-Powered Cargo Fleet", *New Flexport*, December 2, 2015. https://www.flexport.com/blog/nuclear-powered-cargo-ships/. Accessed on October 17, 2021.

not pursued as they were too small, and were expensive to operate. The 1988-built Russian vessel *Sevmorput* is one of only four nuclear-powered merchant ships ever built.[32] After refurbishment in 2016, currently it is the only one in service in the world. The Soviet Union, and now Russia, has been using nuclear-powered icebreaker ships since the late 1950s. A few are still in service[33], and more are being built.

## Atomic Battery: Military Applications

The atomic battery generates electricity essentially using energy released during the decay of radioactive isotopes. Unlike nuclear reactors, atomic batteries do not use a chain reaction, but current is generated by charged particle radiation. They are costly, but have an extremely long life, and are ideal for stand-alone systems and platforms such as pacemakers, spacecraft, underwater systems and equipment left in uninhabited remote places.

Atomic batteries are of two types. The thermal types use the heat emanating from the nuclear isotopes' decay to generate electricity. The second method is to tap the energy directly from the emitted radiation. Thermo-photovoltaic cells are more efficient than thermo-electric ones. The first plutonium-powered pacemaker, the Numec NU-5, was used for a human patient in 1970.[34] Unlike the battery operated pacemaker which requires surgical replacement of batteries every 5 to 10 years, the plutonium powered one can carry on for the whole life.

---

32. Thomas Nilsen, "Sevmorput Returns to Arctic After Troublesome Year", *The Barents Observer*, August 10, 2021. https://thebarentsobserver.com/en/nuclear-safety/2021/08/sevmorput-returns-arctic-after-troublesome-year. Accessed on October 17, 2021.
33. Alexandra Odynova, "Russia Touts Huge New Nuclear-Powered Icebreaker as Proof the Arctic is Ours", CBS News, September 22, 2020. https://www.cbsnews.com/news/russia-touts-huge-new-nuclear-powered-icebreaker-as-proof-the-arctic-is-ours/ Accessed on October 27, 2021.
34. Sam Brusco, "MedTech Memoirs: The Plutonium-Powered Pacemaker, Medical Design and Outsourcing", January 13, 2016. https://www.medicaldesignandoutsourcing.com/medtech-memoirs-the-plutonium-powered-pacemaker/. Accessed on October 28, 2021.

In war zones, when the troops are on the move, electric supply to charge devices could be an issue. Generators require fuel supplies to be carried. Large battery packs are too heavy to carry. The US Defence Science Board considers "nuclear batteries" as essential technology for the US military in the years ahead.[35] Nuclear micro and nano-batteries would also be used to power very small dust-sized, electronics which are called "smart-dust".[36] The tritium nuclear battery can last for around 13 years. Disposal of such batteries will remain an issue. However, considering that the much larger and numerous chemical batteries have much more serious disposal issues, this should be manageable. Nuclear batteries are much smaller and have a much longer life, so the numbers for disposal are relatively low. The nuclear battery isotopes will be unable to undergo nuclear fission, as such, they cannot be used as fuel for a bomb.

## Nuclear Propulsion for Satellites

Nuclear power is being used in space for electricity and heat. Radioisotope thermoelectric generators have been used in long distance space probes and on manned lunar missions. A radioisotope heater unit prevents components from becoming too cold to function. The devices could be powered for decades. Both the USA and USSR have sent many nuclear-electric satellites into space. The more powerful TOPAZ-II reactor[37] could produce 10 Kilowatts (KW) of electricity. Nuclear power for space propulsion systems using ion thrusters greatly reduces the satellite size and increases options for alternative payloads. Satellites drift in space and require propulsion to regain the correct

---

35. Steve Weintz, "The US Military is Working on Nuclear Batteries", Medium.com, November 3, 2014. https://medium.com/war-is-boring/powering-the-militarys-future-batteries-nuclear-style-472975d7de8. Accessed on October 27, 2021.
36. Jamie D. Phillips, "Energy Harvesting in Nanosystems: Powering the Next Generation of the Internet of Things", *Frontiers In*, March 21, 2021. https://www.frontiersin.org/articles/10.3389/fnano.2021.633931/full. Accessed on October 28, 2021.
37. Jack Craddock III, "Nuclear Power in Space: The TOPAZ Reactor", Stanford University, Winter 2016, March 15, 2016. http://large.stanford.edu/courses/2016/ph241/craddock2/. Accessed on October 28, 2021.

allotted satellite slot.[38] The envisaged missions in space would require greater manoeuvrability, and, therefore, lightweight, long-lasting nuclear power solutions will help. In April 2021, the Defence Advanced Research Projects Agency (DARPA) issued contracts for space-based nuclear thermal propulsion systems.[39]

## Way Ahead for India

Clearly, while environmental concerns are driving the switch from fossil fuel power generation to much cleaner alternative energy for electric power generation for civil use, the military needs much smaller, lighter, and long lasting power sources for better mobility and at remote locations. Space is another area with great applications for miniature nuclear power plants. India has a well-established nuclear energy programme. India has the third largest armed forces and very active borders spanning some of the highest mountains and remote jungles. Like those of the other major powers, the Indian armed forces require small nuclear power plants for use on the move. India's nuclear submarine programme is still evolving, and it has still to begin developing a nuclear-powered aircraft carrier. Also, while India's space programme is highly mature, it has still to develop and deploy nuclear batteries in space. The potential is promising and it is time for India to make a clear assessment of the technology and its benefits.

38. Nathan Strout, "Pentagon Taps Industry for Nuclear-Powered Propulsion for its Satellites", c4isrnet, September 11, 2021. https://www.c4isrnet.com/battlefield-tech/space/2021/09/10/pentagon-taps-industry-for-nuclear-powered-propulsion-for-its-satellites/. Accessed on October 27, 2021.
39. Ibid.

# Nuclear Non-proliferation

# 5. Nuclear Doctrines of Great Powers and Vertical Nuclear Proliferation Trends

*Abhishek Saxena*

## Introduction

While the risk of Non-Nuclear Weapon States (NNWS) acquiring nuclear weapons remains, the existing Nuclear Weapon States (NWS) are busy quantitatively and qualitatively modernising their nuclear forces at an accelerated pace. This chapter analyses the nuclear doctrines and vertical nuclear proliferation trends of the NWS recognised by the Treaty on Non-Proliferation of Nuclear Weapons (NPT).

## United States

The United States is the foremost nuclear power having the second-largest nuclear weapons stockpile after Russia. Since the 1960s, the US has maintained a strategic nuclear triad. To preserve deterrence stability and help assure allies and partners, the US has always maintained a nuclear first use doctrine. The primary objective of the US' nuclear strategy is to deter potential adversaries from waging a nuclear attack of any scale on US soil.[1] In addition to that, the US nuclear strategy contributes to deterrence against a "non-nuclear attack, assurance of allies and partners, achievement of US objectives, if deterrence fails, and capacity to hedge against an uncertain future."[2] It also seeks to deter the proliferation of nuclear weapons and prohibit terrorists from

1. US Department of Defence, "Nuclear Posture Review 2018", February 2, 2018, p. 20, https://media.defence.gov/2018/Feb/02/2001872886/-1/-1/1/2018-NUCLEAR-POSTURE-REVIEW-FINAL-REPORT.PDF. Accessed on October 17, 2021.
2. Ibid., p. 20.

accessing nuclear technology and radioactive material. According to the US declaratory policy, it would use nuclear weapons only under "extreme circumstances" to "defend the vital interests of the United States, its allies, and partners".[3] In addition to a strategic attack, extreme circumstances could include a "non-nuclear strategic attack", a significant attack on the US or allied population centres, infrastructure, nuclear command and control and attack assessment capabilities.[4] The 2018 Nuclear Posture Review (NPR) reiterates the US' stand on maintaining a first use doctrine. It mentions that "to help preserve deterrence and the assurance of allies and partners, the United States has never adopted a 'no first use' policy and, given the contemporary threat environment, such a policy is not justified today."[5]

Currently, the US strategic triad consists of strategic bombers equipped with gravity bombs and Air-Launched Cruise Missiles (ALCMs), nuclear ballistic missile submarines (SSBNs) carrying Submarine-Launched Ballistic Missiles (SLBMs) and land-based Inter-continental Ballistic Missiles (ICBMs). The 2018 NPR justified the continuation of the nuclear triad to enhance deterrence survivability against a nuclear first strike and to strengthen deterrence stability by having the capacity to hold multiple adversary targets at risk.[6] It maintains, "The triad's synergy and overlapping attributes help ensure the enduring survivability of our deterrence capabilities against attack and our capacity to hold a range of adversary targets at risk throughout a crisis or conflict. Eliminating any leg of the triad would greatly ease adversary attack planning and allow an adversary to concentrate resources and attention on defeating the remaining two legs."[7] The US strategic triad's three legs include a fleet of 14 nuclear-capable Ohio-class submarines equipped with 20 Trident II D5 missile systems, an

3. Ibid.
4. Ibid.
5. Ibid., p. 22.
6. Ground-based ICBMs provide a prompt and responsive strategic strike option, SSBNs provide a survivable retaliatory capability, and air-based deterrence offers visibility through the positioning of forces to signal national resolve during a crisis.
7. n. 1, p. 43.

underground silo-based ICBM force of 400 single-warhead Minuteman III ICBMs, and a bomber force of 46 nuclear-capable B-52H bombers, and 20 nuclear-capable B-2A stealth strategic bombers.[8]

Since the Cold War's conclusion, both as a part of its nuclear modernisation efforts and in response to various arms control treaties, the United States has drastically reduced its strategic nuclear weapon capability along all three legs of the nuclear triad. Many of the weapon systems in the US strategic forces are more than three decades old, having survived beyond their designated service life through multiple upgrades and life extension programmes. However, the modernisation of legacy nuclear systems and their prospective replacement in the mid-2020s or early 2030s is the utmost priority of the Department of Defence (DoD) and Department of Energy (DoE).[9]

The United States first deployed the Minuteman III ICBM in 1970, and it will be finally withdrawn from service after six decades in 2030, making it the world's oldest deployed strategic missile in the world. As a replacement to the Minuteman ICBMs, the United States is developing the Ground-Based Strategic Deterrent (GBSD), a more capable and modern missile delivery system, planned to be deployed in 2029.[10] The new ICBM, when deployed, would carry an upgraded and life-extended version of the W78 warhead currently carried by Minuteman III missiles.

Trident submarines, first deployed in 1971, had an original service life of 30 years, later extended to 42 years. The submarines in this class would start retiring from operational capacity in the late 2020s. As a replacement for the Ohio class submarines, the United States is coming up with a fleet of 12 new nuclear-capable ballistic missile submarines, known as the Columbia class, expected to join the US Navy in 2031. During their initial operational life till 2039, the Columbia class submarines will carry the life-extended version of the Trident II D5 missiles. The future requirement of

---

8. Hans M. Kristensen and Matt Korda, "United States Nuclear Weapons, 2021", *The Bulletin of the Atomic Scientists*, Vol. 77, No. 1, 2021, pp. 43-63.
9. n. 1, p. 48.
10. Amy F. Woolf, *US Strategic Nuclear Forces: Background, Developments, and Issues*, RL33640 (Washington DC: Congressional Research Service, 2021), pp. 17-21.

the submarines in class for the complete expected service life (till 2084) would be met by doubly life extended and upgraded Trident missiles being developed under the D5 Life Extension II (D5LE2) programme. Also, the National Nuclear Security Administration (NNSA) is working on a new sea-based warhead named W93, which will eventually replace the currently deployed W76 and W78 warheads on the life-extended Trident missiles.[11]

Since the Cold War, the B-2 and B-52 heavy bombers have continued to serve the US Air Force in both conventional and nuclear roles. They have undergone various upgrades to sustain their capabilities but may not meet the emerging challenges. As a result, to maintain the heavy bombers' contribution to the US strategic triad, the US is developing a new strategic bomber, known as the B-21 Raider. It is a manned, subsonic aircraft with some stealth capabilities. The US Air Force plans to induct between 80 and 100 new bombers, scheduled to enter service in 2025.[12] It is also developing a new advanced long-range stand-off cruise missile to replace the ageing ALCM deployed on B-52H bombers.[13]

Washington is increasingly concerned about what US strategists call the Russian "escalate to de-escalate" doctrine—the Russian perception that the threat of limited nuclear escalation or the potential use of a non-strategic nuclear weapon can provide it with a coercive advantage over the United States in a state of crisis or conflict.[14] The United States intends to expand its non-strategic weapons capability to correct the Russian strategic misperception in the advantageous application of its tactical nuclear weapons. In contrast to the conventional argument that flexible nuclear options would lower the nuclear threshold and enable nuclear war-fighting, the US argues that "expanding US tailored response options will raise the nuclear threshold and help ensure that potential adversaries perceive no possible advantage in limited nuclear escalation,

11. Ibid., pp. 32-33.
12. Ibid., p. 48.
13. The first missile is slated for completion in 2026 and will replace the ALCM after 2030.
14. n. 1, pp. 52-53.

making nuclear weapons employment less likely."[15] The United States is modifying some existing SLBM warheads into low yield nuclear-armed weapons to enhance its immediate war-fighting ability. In the longer term, it would develop nuclear-armed Sea-Launched Cruise Missiles (SLCMs).[16]

The NPR claims that the US nuclear posture will not remain stagnant and will evolve with the international environment, the emergence of geopolitical and technological threats and the modernisation of nuclear capability. In the months leading up to the Biden Administration's Nuclear Posture Review, the United States, yet again, is debating whether to adopt a No First Use (NFU) nuclear posture. There are underlying reports that President Biden is considering to declare that "the sole purpose of the US nuclear arsenal is to deter or retaliate against a nuclear attack."[17] Whether such a radical shift would be made is entirely at the discretion of the president and his close associates at the White House. However, given the current international security environment, the revelations about China's expansive silo construction and hypersonic tests, the emerging need to deter two nuclear adversaries, and, most importantly, the extended deterrence commitments and associated nuclear proliferation risks, the United States, in the author's view, is less likely to adopt an NFU posture.

### Russia

The Soviet nuclear doctrine during the Cold War evolved with the changing dynamics of the US-Soviet relationship and furtherance of arms control agreements. A leading Sovietologist argues that for the first quarter-century of the nuclear age, "the fundamental assumption of the Soviet military doctrine was that, if a global war was unleashed

---

15. Ibid., p. 54.
16. This is in addition to the existing US tactical weapons capability of the B61-3, 4, and 10 gravity bombs deployed on the F-15E dual capability aircraft and upcoming B61-12 gravity bomb.
17. Paul Sonne and John Hudson, "Biden Administration Considers Adjusting Rationale for US Nuclear Arsenal", *The Washington Post*, November 2, 2021. https://www.washingtonpost.com/national-security/biden-nuclear-weapons-strategy/2021/11/02/story.html. Accessed on November 12, 2021.

by the 'imperialist West', the Soviet Union could defeat the enemy and achieve victory, despite the enormous ensuing damage."[18] However, with the signing of the Anti- Ballistic Missile (ABM) Treaty in 1972, the Soviet Union and the US accepted the condition of mutual vulnerability in which both sides had nuclear capabilities to inflict unacceptable damage, where defence was not an option.[19] In 1982, Soviet General Secretary Leonid Brezhnev announced the Soviet no first use doctrine, giving up the option of using nuclear weapons first.[20]

Since the end of the Cold War, the Russian reliance on nuclear weapons has increased. To counter the weakness in the conventional realm, in 1993, Russia explicitly rejected the Soviet Union's NFU doctrine. The 2000 Russian military doctrine called for the preemptive use of nuclear weapons in response to "large scale aggression utilising conventional weapons in situations critical to the national security of the Russian Federation."[21] The 2010 military doctrine curtailed the conditions under which Russia might contemplate using nuclear weapons. Whereas the previous doctrine allowed for nuclear weapons use when the "national security of the Soviet Union is under threat," the 2010 doctrine limited the use of nuclear weapons to the situation when the state's very existence is under threat.[22] The most recent military doctrine published in 2014 restated the position on nuclear weapons use under the conventional survival threat. It maintains that the Russian Federation "shall reserve the right to use nuclear weapons in response to the use of nuclear and other types of weapons of mass destruction against it or its allies, as well as in the event of aggression against the Russian

18. Alexey Arbatov, "Understanding the US-Russia Nuclear Schism", *Survival*, Vol. 59, No. 2, 2017, p. 39.
19. Andrei Shoumikhin, "Nuclear Weapons in Russian Strategy and Doctrine", in Stephen J. Blank, ed., *Russian Nuclear Weapons: Past, Present, and Future*, (Pennsylvania: Strategic Studies Institute), p. 104.
20. Ibid., p. 106.
21. Nikolai Sokov, "Russia's 2000 Military Doctrine," *Nuclear Threat Initiative*, September 30, 1999. https://www.nti.org/analysis/articles/russias-2000-military-doctrine/. Accessed on October 4, 2021.
22. The text of the 2010 Russian Military Doctrine is available at: https://carnegieendowment.org/files/2010russia_military_doctrine.pdf, p. 9.

Federation with the use of conventional weapons when the very existence of the state is in jeopardy."[23] Russian President Vladimir Putin restated the above statement while delivering the 2018 presidential address to the Federal Assembly.[24] The subsequent Russian documents, including the one released in June 2020 titled "On Basic Principles of State Policy of the Russian Federation on Nuclear Deterrence," have retained similar language on the case of nuclear weapons, not explicitly stating the use of atomic weapons in the use of a conventional attack, but maintaining an ambiguity regarding nuclear escalation.

The US projection of its modern and precise conventional capabilities during the Gulf War and Kosovo War raised Russian concerns about American intervention on behalf of the Chechnya rebels. Thus, to deter possible American intervention, Russia, in its 2000 military doctrine, envisioned threatening an adversary with limited nuclear weapons, forcing him to back down and maintain the status quo. Western analysts have dubbed the Russian tailored nuclear strategy to deter Western involvement in conflicts critical to Russian security as a "de-escalation strategy".[25] Since then, in all its major military exercises, Russia has conducted simulations for a limited nuclear strike at all possible ranges.

There is an underlying debate over whether Russia has adopted an "escalate to de-escalate" doctrine, a term of Western creation and missing from Russian White Papers. The 2018 Nuclear Posture Review of the US Department of Defence expressed grave concern about the Russian "escalate to de-escalate" strategy and recommended developing low-yield tactical nuclear warheads.[26] It also mentioned explicitly that the

23. See the text of the 2014 Russian Military Doctrine at: https://rusemb.org.uk/press/2029.
24. The full official English transcript of the 2018 Presidential Address to the Federal Assembly is available at: http://en.kremlin.ru/events/president/news/56957.
25. Nikolai N. Sokov, "Why Russia Calls a Limited Nuclear Strike 'De-escalation'", *The Bulletin of the Atomic Scientists,* March 13, 2014. https://thebulletin.org/2014/03/why-russia-calls-a-limited-nuclear-strike-de-escalation/. Accessed on October 12, 2021.
26. n. 1, pp. 30-55.

Russian use of non-strategic nuclear weapons, "however limited, will fail to achieve its objectives, fundamentally alter the nature of a conflict, and trigger incalculable and intolerable costs for Moscow."[27] Some of the experts, closely observing the Russian military and simulation exercises, concur with the US government's assessment of the Russian nuclear doctrine.[28] On the contrary, those studying the government documents and White Papers argue that the evidence of Russia employing nuclear weapons at the lower level of conflict is weak.[29] The Russian rationale to include nuclear use against conventional aggression was a stop-gap arrangement until Russia built its conventional strike capability. Recent RAND reports have pointed out that since 2000, Russia has modernised its conventional strike capability.[30] Scholars postulate that Russian reliance on nuclear weapons has decreased with enhanced conventional deterrence, thus, enhancing its nuclear threshold.[31]

Like the United States, Russia maintains a triad of strategic nuclear forces consisting of 318 ICBM systems, ten ballistic missile submarines of the Delta IV, Delta III, and Borei classes, and two divisions of Tu-160 Blackjack and Tu-95 MS Bear H heavy bombers.[32] Russia is developing a new advanced silo-based ICBM, the RS-28 Sarmat, which is expected

27. Ibid., p. 30.
28. Matthew Kroenig, "The Case for Tactical US Nukes", *Wall Street Journal,* January 24, 2018. https://www.wsj.com/articles/the-case-for-tactical-u-s-nukes-1516836395. Accessed on October 14, 2021.
29. Olga Oliker and Andrey Baklitsky, "The Nuclear Posture Review and Russian 'De-Escalation:' A Dangerous Solution to a Non-Existent Problem", *War on the Rocks,* February 20, 2018. https://warontherocks.com/2018/02/nuclear-posture-review-russian-de-escalation-dangerous-solution-nonexistent-problem/. Accessed on October 15, 2021.
30. Scott Boston and Dara Massicot, "The Russian Way of Warfare", RAND Corporation, 2017. https://www.rand.org/pubs/perspectives/PE231.html. Accessed on October 16, 2021.
31. Kristin Ven Bruusgaard, "The Myth of Russia's Lowered Nuclear Threshold", *War on the Rocks,* September 22, 2017. https://warontherocks.com/2017/09/the-myth-of-russias-lowered-nuclear-threshold/. Accessed on October 15, 2021.
32. Kristensen and Korda, n. 8, pp. 73-84.

to enter combat duty by late 2022.[33] In addition to three existing Borei class submarines, Russia is constructing five more Borei class SSBNs, planned to be completed by 2023, and eventually plans to deploy 10 Borei class submarines by 2027.[34] Russia is also planning to field a newer, stealthier, longer-range version of the Tu-160, known as the Tu-160M2, manufacturing of which will start in the mid-2020s.[35] While delivering the 2018 state of the union address, President Putin unleashed a new generation of nuclear weapons, including what he described as an "invincible" global range cruise missile, an "unmanned" underwater vehicle, and a heavy intercontinental missile with "practically no range restrictions".[36] In addition to modernising its strategic forces, Russia is leading the race on hypersonic weapons. It deployed its first regiment of hypersonic missile systems named Avangard in December 2019.[37] Lately, in October 2021, Russia successfully fired the ship-launched Tsirkon hypersonic missile from the *Severodvinsk* submarine.[38] Earlier,

33. Peter Suciu, "RS-28 Sarmat: New Russian ICBM to Enter Combat Duty in 2022", *The National Interest,* December 17, 2020. https://nationalinterest.org/blog/buzz/rs-28-sarmat-new-russian-icbm-enter-combat-duty-2022-174616. Accessed on October 8, 2021.
34. n. 10, pp. 16-17.
35. Mark Episkopos, "Russia Has Big Plans for Its 'New' Tu-160 Supersonic Bombers", *The National Interest*, June 9, 2021. https://nationalinterest.org/blog/buzz/russia-has-big-plans-its-new-tu-160-supersonic-bombers-187278. Accessed on October 8, 2021.
36. The official English transcript of the Presidential Address to the Federal Assembly delivered at the Manezh Central Exhibition Hall on March 1, 2018, can be found at: http://en.kremlin.ru/events/president/news/56957. Accessed on November 25, 2021.
37. "First Regiment of Avangard Hypersonic Missile Systems Goes on Combat Duty in Russia", *TASS,* December 27, 2019. https://tass.com/defence/1104297. Accessed on August 1, 2021.
38. "Russia Test-Fires new Hypersonic Missile from Submarine", *The Hindu*, October 4, 2021. https://www.thehindu.com/news/international/russia-test-fires-new-hypersonic-missile-from-submarine/article36820524.ece. Accessed on October 5, 2021.

in July 2021, the Tsirkon missile was successfully test-launched from the *Admiral Gorshkov* frigate.[39]

While the Russian Federation has aggressively modernised and diversified its nuclear force posture in the past decade, its nuclear doctrine has remained essentially unchanged. An array of factors such as the emerging international security environment, prospects of the ongoing US-Russia strategic stability dialogue, and Russia's relations with the West would determine the future trajectory of the Russian nuclear posture.

## China

The proclaimed Chinese nuclear doctrine has remained unchanged since Chairman Mao Zedong first articulated it.[40] China's political elite had an inherent belief in the effectiveness of a small survivable nuclear force to inflict unacceptable damage in a retaliatory strike, thus, deterring a nuclear attack.[41] In the statement released by the government of the People's Republic of China (PRC) after the successful conduct of its first atomic bomb test, Beijing declared that "China will never at any time and under any circumstances be the first to use nuclear weapons."[42] A few days later, Mao observed in an interview with Edgar Snow, "We don't wish to have too many atom bombs ourselves. What

39. Ewan Palmer, "Russia Tests Hypersonic Missile, Hits Target More Than 200 Miles Away", *Newsweek,* July, 19, 2021. https://www.newsweek.com/russia-hypersonic-cruise-missile-launch-1611020. Accessed on November 2, 2021.
40. It is because China's top Party leaders have dominated and controlled the nuclear weapons strategy; they never delegated authority over nuclear strategy to seniors military leaders. China's nuclear strategy reflects ideas forcefully articulated by Mao Zedong, Zhou Enlai, and Deng Xiaoping, and receptively espoused by military leaders of the Second Artillery.
41. M. Taylor Fravel, *Active Defence: China's Military Strategy Since 1949* (Princeton, New Jersey: Princeton University Press, 2019), p. 238.
42. "Statement of the Government of the People's Republic of China," *History and Public Policy Program Digital Archive*, October 16, 1964. https://digitalarchive.wilsoncenter.org/document/134359.pdf?v=b1e04ac05705. Accessed on October 25, 2021.

would we do with so many? To have a few is just fine".[43] The complete treatment of China's nuclear strategy has been articulated in the 2006 Defence White Paper. The paper stated that China pursues a "self-defensive nuclear strategy", whose enduring principles are "counter-attack in self-defence" and "limited development" of nuclear weapons. The most recent Chinese Defence White Paper[44], released in July 2019, has reiterated the enduring aspects of China's proclaimed nuclear doctrine:

> China is always committed to a nuclear policy of no first use of nuclear weapons at any time and under any circumstances, and not using or threatening to use nuclear weapons against non-nuclear-weapon states or nuclear weapon-free zones unconditionally... China does not engage in any nuclear arms race with any other country and keeps its nuclear capabilities at the minimum level required for national security. China pursues a nuclear strategy of self-defence, the goal of which is to maintain national strategic security by deterring other countries from using or threatening to use nuclear weapons against China.

China's nuclear doctrine of no first use of nuclear weapons preordains it towards an assured retaliation strategy[45]: surviving the nuclear first strike and then launching a massive retaliatory counter-strike. China has never looked for numerical equality with its adversaries and, with its nuclear strategy of assured retaliation, has sought to deter nuclear attack and nuclear coercion.

---

43. Quoted in Fravel, n. 41, p. 239.
44. The State Council Information Office of the People's Republic of China, *China's National Defence in a New Era*, July 2019. http://www.xinhuanet.com/english/download/whitepaperonnationaldefenseinnewera.doc. Accessed on October 24, 2021.
45. M. Taylor Fravel and Evan S. Medeiros, "China's Search for Assured Retaliation: The Evolution of Chinese Nuclear Strategy and Force Structure," *International Security*, Vol. 35, No. 2, 2010, pp. 48-87.

China's nuclear forces comprise a well-established land-based and sea-based deterrent and a developing air-based deterrent. Its strategic triad includes multiple land-based missiles of varying ranges and warhead capability, the latest being the Dongfeng-17, a solid-fuelled road-mobile Medium Range Ballistic Missile (MRBM), and Dongfeng-41, a solid-fuelled road-mobile ICBM; six Jin class ballistic missile submarines armed with JL-2 SLBMs; and H-6 bomber aircraft.[46] China has maintained a low-alert level for its nuclear posture, as was highlighted by the recent Pentagon report, "China almost certainly keeps the majority of its nuclear force on a peace-time status—with separated launchers, missiles, and warheads."[47]

Starting in the early 1980s and moving on to an accelerated pace in the 1990s and 2000s, China developed its nuclear forces. The development efforts aim to replace the liquid-fuelled, slow-launching and shorter-range missiles with solid-fuelled, quicker-launching, longer-range, and road-mobile missile systems. During this period, some upgraded or newly introduced missiles are the DF-31, the first solid-fuelled road-mobile ICBM introduced in 2006[48]; solid-fuelled DF-41 ICBM; dual-capable DF-26 Intermediate Range Ballistic Missile (IRBM); and DF-17 hypersonic boost-glide missile.[49] In addition to six Jin class submarines, China plans to construct a quieter third-generation (Type 096) SSBN, armed with the JL-3 SLBM, with a speculated range of 9,000 km. In the last couple of years, China has reassigned its nuclear role to the People's Liberation Army Air Force (PLAAF). To extend the range of its H-6 bomber aircraft, China is developing a new long-range stealth

46. Kristensen and Korda, n. 8, pp. 443-457.
47. Office of the Secretary of Defence, "Annual Report to Congress: Military and Security Developments Involving the People's Republic of China", September 1, 2020, p. 8. https://media.defence.gov/2020/Sep/01/2002488689/-1/-1/1/2020-DOD-CHINA-MILITARY-POWER-REPORT-FINAL.PDF. Accessed on October 22, 2021.
48. Subsequently, China has introduced an upgraded versions of the DF-31 such as DF-31A and DF-31AG.
49. Kristensen and Korda, n. 8, pp. 447-451.

bomber, known as the H-20.[50] Also, China is in the advanced stage of operationalising its first Air-Launched Ballistic Missile (ALBM), having tested the missile twice in 2016 and 2018.

*The Bulletin of the Atomic Scientists*, in its recent Nuclear Notebook, estimates that "China has produced a stockpile of approximately 350 nuclear warheads for delivery by approximately 280 operational land-based ballistic missiles, 72 sea-based ballistic missiles, and 20 nuclear gravity bombs assigned to bombers."[51] This is a higher estimate as compared to the "low-200s" reported by the Pentagon's 2020 report to Congress.[52] In an alarming revelation, the Pentagon's 2021 report forecasts that the continued expansion, diversification, and modernisation of its nuclear forces might enable the People's Republic of China (PRC) "to have up to 700 deliverable nuclear warheads by 2027" and at least 1,000 warheads by 2030.[53]

The sense of alarm sounded by the Pentagon's 2021 report follows the recent discovery and revelation about China's massive silo construction and hypersonic missile tests. In recent months, three US-based researchers, using Open-Source Intelligence (OSINT), discovered three under construction missile silo fields in China located at Yumen[54],

---

50. Ibid., p. 453.
51. Ibid., pp. 318-336.
52. US Department of Defence, "Annual Report to Congress: Military and Security Developments Involving the People's Republic of China", p. 85, September 1, 2020. https://media.defence.gov/2020/Sep/01/2002488689/-1/-1/1/2020-DOD-CHINA-MILITARY-POWER-REPORT-FINAL.PDF. Accessed on October 20, 2021.
53. US Department of Defence, "Military and Security Developments Involving the People's Republic of China 2021", p. 92, November 3, 2021. https://media.defence.gov/2021/Nov/03/2002885874/-1/-1/0/2021-CMPR-FINAL.PDF. Accessed on November 8, 2021.
54. Joby Warrick, "China is Building More than 100 new Missile Silos in its Western Desert, Analysts Say", *The Washington Post*, June 30, 2021, https://www.washingtonpost.com/national-security/china-nuclear-missile-silos/2021/06/30/0fa8debc-d9c2-11eb-bb9e-70fda8c37057_story.html. Accessed on July 4, 2021.

Hami[55] and Hanggin Banner.[56] That would expectedly add 250-300 new missile systems to the Chinese arsenal. Lately, in a brilliant scoop, the *Financial Times* indicated that China conducted two hypersonic missile tests on July 27 and August 13, 2021, respectively.[57] The nuclear-capable hypersonic missile tested in July "circled the globe before speeding towards its target, demonstrating an advanced space capability that caught US intelligence by surprise."[58]

Recent developments have triggered speculation about a possible shift in China's nuclear doctrine. Scholars argue that expansive silo construction and hypersonic missile capability may allow China to shift from a low-alert nuclear retaliatory posture towards a high-alert launch on warning posture.[59] However, the Western alarm about the Chinese nuclear developments is overstated. First, the number of under-construction missile silos is not equivalent to the number of deployed

55. Hans Kristensen and Matt Korda, "China Is Building a Second Nuclear Missile Silo Field", *Federation of American Scientists*, July 26, 2021. https://fas.org/blogs/security/2021/07/china-is-building-a-second-nuclear-missile-silo-field/. Accessed on August 17, 2021.
56. Rod Lee, "PLA Likely Begins Construction of an Intercontinental Ballistic Missile Silo Site near Hanggin Banner", China Aerospace Studies Institute, August 12, 2021, https://www.airuniversity.af.edu/CASI/Display/Article/2729781/pla-likely-begins-construction-of-an-intercontinental-ballistic-missile-silo-si/. Accessed on August 20, 2021.
57. Demetri Sevastopulo, "China Conducted Two Hypersonic Weapons Tests This Summer", *Financial Times,* October 20, 2021;. https://www.ft.com/content/c7139a23-1271-43ae-975b-9b632330130b. Accessed on October 27, 2021.
58. Demetri Sevastopulo, "China Tests New Space Capability with Hypersonic Missile" *Financial Times*, October 17, 2021. https://www.ft.com/content/ba0a3cde-719b-4040-93cb-a486e1f843fb. Accessed on October 19, 2021.
59. Hans Kristensen and Matt Korda, "China's Nuclear Missile Silo Expansion: From Minimum Deterrence to Medium Deterrence", *The Bulletin of the Atomic Scientists,* September 1, 2021. https://thebulletin.org/2021/09/chinas-nuclear-missile-silo-expansion-from-minimum-deterrence-to-medium-deterrence/. Accessed on September 4, 2021; Suyash Desai, "An Expert Explains: Why China Seems to be Building Three Missile Silos", *Indian Express,* September 2, 2021. https://indianexpress.com/article/explained/china-missile-silo-test-nuclear-weapons-7471044/. Accessed on September 8, 2021.

missiles; "China might build more silos than missiles to create a 'shell game' that would make it harder for an adversary to target the missiles."[60] Second, the expansive silo construction might be a Chinese reaction to the US' strengthening its missile defence, which China perceives as a threat to the survivability of its retaliatory nuclear force. Third, China's hypersonic missile tests perhaps demonstrate its capability to penetrate advanced US missile defences, entrenching the US in a deeper state of mutual nuclear vulnerability. Fourth, despite the changes in force posture, China's nuclear declaratory policy emphasises the no first use of nuclear weapons, retaliatory nuclear weapons capability, and a survivable nuclear force. In essence, China's nuclear modernisation and expansion efforts seek survivability for its nuclear forces and the ability to penetrate US missile defences.

### United Kingdom

The UK was the third country after the US and Russia to develop nuclear weapons. It exploded its first nuclear bomb in the hull of the HMS *Plym*, a Royal Navy frigate, off the coast of Monte Bello Islands, western Australia, on October 3, 1952.[61] At the peak of the Cold War (the late 1970s), the UK had around 520 nuclear warheads. In the past three decades, the UK has gradually reduced its nuclear stockpile and eased its nuclear posture. The 2010 Strategic Defence and Security Review (SDSR) released by the Conservative government announced an intention to reduce the overall nuclear stockpile from 225 to 180 warheads by the mid-2020s.[62] However, in the recently

60. Hans M. Kristensen and Matt Korda, "Chinese Nuclear Weapons, 2021," *The Bulletin of the Atomic Scientists,* Vol. 77, No. 6, pp. 318-336. https://www.tandfonline.com/doi/full/10.1080/00963402.2021.1989208. Accessed on November 22, 2021.
61. Claire Taylor, Tim Youngs, Ross Young, and Gavin Berman, *The Future of the British Nuclear Deterrent*, Research Paper 06/53 November 3, 2006. (London: House of Commons Library, 2006), p. 9.
62. H.M. Government, "Securing Britain in an Age of Uncertainty: The Strategic Defence and Security Review", pp. 37-39, October 2010. https://assets.publishing.service.gov.uk/government/uploads/system/uploads/attachment_data/file/62482/strategic-defence-security-review.pdf, accessed on July 12, 2021.

released Integrated Review of Security, Defence, Development and Foreign Policy, the UK announced that it no longer intends to reduce its nuclear stockpile; contrary to that, the UK has decided to increase its nuclear warheads from 225 to 260 warheads.[63]

The UK has a nuclear doctrine of a minimum, credible, and independent nuclear deterrent. The objective of its nuclear strategy is to "preserve peace, prevent coercion and deter aggression."[64] In addition to previously identified risks to the UK from major nuclear-armed states, emerging nuclear states, and state-sponsored nuclear terrorism, the Integrated Review has expressed concern about some states "significantly increasing and diversifying their nuclear arsenals."[65] It indicates that they are "investing in novel nuclear technologies and developing new 'war-fighting' nuclear systems which they are integrating into their military strategies and doctrines and into their political rhetoric to seek to coerce others."[66] To address these emerging technological and doctrinal challenges, the UK has decided to increase its nuclear arsenal and has extended its long-standing policy of deliberate ambiguity, and would "no longer give public figures for its operational stockpile, deployed warhead or deployed missile numbers."

In addition to preserving British security, the UK has earmarked its nuclear forces for the collective security of the Euro-Atlantic area through the North Atlantic Treaty Organisation (NATO). The 2020 Integrated Review mentions that the UK would "consider using nuclear weapons only in extreme circumstances of self-defence, including the defence of NATO Allies".[67] In a negative security assurance, the UK

63. H.M. Government, "Global Britain in a Competitive Age: The Integrated Review of Security, Defence, Development and Foreign Policy," p. 76, March 16, 2021. https://www.gov.uk/government/collections/the-integrated-review-2021. Accessed on July 4, 2021.
64. Ibid., p. 76.
65. Ibid.
66. Ibid.
67. Ibid.

assures the non-nuclear weapon states party to the NPT of non-use of nuclear weapons and nuclear threats against them.

The British nuclear deterrent is sustained by a fleet of four nuclear-armed ballistic missile submarines—the Vanguard class submarines—armed with the Trident II D5 missiles carrying Mk4/A warheads.[68] Since April 1969, one of the British submarines has always remained on a Continuous At-Sea Deterrent (CASD), patrolling the sea waters worldwide undetected. The 2021 Integrated Review has restated the British resolve to maintain a submarine-based deterrent and relentless underwater sea patrols. The British submarines on patrol are at "several days' notice to fire", and since 1994, missiles on-board deterrent submarines do not target any particular state.

The flagship of the UK's nuclear forces, the Vanguard class submarines, joined the Royal Navy in December 1994 and will be phased out in the early 2030s.[69] Initially, the submarines in the class had a design life of 25 years. It has now been extended to a period of 37-38 years. In March 2007, the British Parliament voted in favour to retain the British strategic deterrent beyond the life of the Vanguard submarines. As a replacement to the current system, the UK is building four new nuclear-armed ballistic missile submarines, known as the Dreadnought class submarines.[70] The first of the four submarines, the *Dreadnought*, is expected to enter service in the early 2030s and cater to the UK's nuclear deterrence requirements until the early 2060s. The UK is participating in the US D-5 Life Extension (D5LE) programme, which will extend the life of the Trident missiles to 2039, catering for the initial operational requirements of the Dreadnought submarines.[71] Also, under the US

68. Hans M. Kristensen and Matt Korda, "United Kingdom Nuclear Weapons, 2021", *The Bulletin of the Atomic Scientists*, Vol. 77, No. 3, 2021, pp. 153-158.
69. UK Ministry of Defence and Foreign and Commonwealth Office, "The History of the UK's Nuclear Weapons Programme", p. 3.
70. Claire Mills, *Replacing the UK's Strategic Nuclear Deterrent: Progress of the Dreadnought Class* (London: House of Commons Library, 2021), pp. 1-24.
71. Woolf, n. 10, pp. 27-28.

W93 warhead programme, Britain is working closely with the American counterparts on its successor warhead to the Mk4/A.[72]

The UK is greatly alarmed about the advances in hypersonic technology and integrated air defences made by Russia and China. The latest Defence White Paper titled *Defence in a Competitive Age* states that the "development of Hypersonic Glide Vehicles… will allow very little warning time and pose significant challenges for defensive missile systems to counter."[73] It also expresses grave concern about the early warning radars and integrated air defence systems which will "enable states to contest and even dominate air space in many areas where the UK will need to operate."[74] To counter the unsettling consequences of hypersonics and to have a way around the Russian missile defences, the Integrated Review has allocated £6.6 billion for novel military research, including hypersonic and laser weapons.[75]

In the light of the unforeseen technological and geopolitical challenges, the UK mentions in its latest Integrated Review that it will keep its nuclear posture under constant review. The prospects of US-Russia arms control agreements, the international security environment, and a possible entry of China into the arms control regime would shape the future trajectory of the UK's nuclear modernisation and vertical proliferation.

## France

The French idea of nuclear deterrence is one of absolute deterrence, as was synthesised by General Charles de Gaulle, founder of the Fifth Republic. He believed that nuclear weapons had revolutionised the

---

72. US Senate Committee on Armed Services, "Statement of Charles A. Richard Commander United States Strategic Command before the Senate Committee on Armed Services", p. 13. https://www.armed-services.senate.gov/download/richard-testimony-042021. Accessed on July 18, 2021.
73. Ministry of Defence, "Defence in a Competitive Age", p. 9, March 22, 2021. https://www.gov.uk/government/publications/defence-in-a-competitive-age. Accessed on July 25, 2021.
74. Ibid.
75. n. 63.

equations of power; a small country with a few kilotons of explosive nuclear material could deter a great power with disproportionate military force.[76] He summed up his thinking on the absolute nature of nuclear deterrence in a Press conference in July 1964: "No doubt, the megatons that we could launch would not equal in number those that the Americans and Russians are able to unleash. But, from a certain nuclear capacity and as far as the direct defence of each one is concerned, the proportion of the respective means no longer has absolute value."[77] Thus, France has always sought "strict sufficiency" of nuclear deterrence or minimum deterrence to assure massive nuclear retaliation to inflict unacceptable damage on the adversary, not numerical parity. France reduced its nuclear warheads arsenal from about 540 warheads at the end of the Cold War to approximately 300 warheads.[78]

For France, nuclear weapons are not just about deterring a strategic or non-strategic attack. According to the 2013 White Paper on Defence and National Security, nuclear deterrence constitutes the ultimate guarantee of French sovereignty.[79] It underlines its ability to take an independent course of action concerning adversaries and alliance members. It is of utmost importance to "the security, protection and independence of the nation".[80] The underlying encompassing rationale of French nuclear deterrence was underlined by President François Hollande in 2015 while addressing the air warriors. He stated, "Deterrence allows us to preserve our freedom of action and decision in all circumstances because it is it

---

76. Bruno Tertrais, "French Nuclear Deterrence Policy, Forces and Future: A Handbook", *Fondation pour la Recherche Stratégique*, February 2020. https://uca.edu/politicalscience/files/2021/10/French-Nuclear-Deterrence-Policy-Forces-and-Future-February-2020.pdf. Accessed on October 27, 2021.
77. Ibid., p. 15.
78. Hans Kristensen and Matt Korda, "French Nuclear Forces", *The Bulletin of the Atomic Scientists*, Vol. 75, No. 1, 2019, p. 1.
79. French Ministry of Defence, "White Paper on Defence and National Security 2013", p. 20, April 29, 2013. https://www.defence.gouv.fr/content/download/215253/2394121/file/White2013.pdf. Accessed on October 30, 2021.
80. Ibid., p.73.

that allows me to avoid any threat of blackmail of state origin that would aim to paralyse us".[81]

The French nuclear doctrine seeks to deter any form of aggression against the state's vital interests.[82] What exactly constitutes vital interests is the president's discretion; however, the official texts, in general, have referred to constituents of the state such as the exercise of sovereignty, protection of the population, and integrity of national territory while referring to vital interests.[83] France does not have a declaratory no first use policy and reserves the right to conduct a limited nuclear strike to demonstrate its resolve and reestablish deterrence. President Emmanuel Macron stated in a speech on defence and nuclear deterrence, "Should there be any misunderstanding about France's determination to protect its vital interests, a unique and one-time-only nuclear warning could be issued to the aggressor state to clearly demonstrate that the nature of the conflict has changed and to reestablish deterrence."[84] Although France is a member of NATO and participates in its nuclear planning mechanisms, it is not a part of its integrated military command structure; its nuclear forces are not formally assigned to NATO.[85]

French nuclear forces during the Cold War comprised a strategic nuclear triad. However, due to the disappearance of the Soviet threat and improvement in the strategic environment in the post-Cold War period, France dismantled its land-based strategic forces based at Plateau d'Albion in 1996.[86] Thus, presently, the French nuclear forces comprise the air and sea-based nuclear deterrent. The French Strategic Ocean Force (FOST) operates four Triomphant class nuclear-powered

81. Statement by Mr. François Hollande, president of the French Republic, on nuclear deterrence, in Istres, on February 19, 2015, is available at: https://www-vie--publique-fr.translate.goog/discours/193954-declaration-de-m-francois-hollande-president-de-la-republique-sur-la?.
82. n. 79, p. 72.
83. Tertrais, n. 76, pp. 25-26.
84. Speech of the President of the Republic on the Defence and Deterrence Strategy, February 7, 2020. https://www.elysee.fr/en/emmanuel-macron/2020/02/07/speech-of-the-president-of-the-republic-on-the-defence-and-deterrence-strategy.
85. Mills, n. 70, p. 3.
86. Ibid.

ballistic submarines, each capable of carrying 16 M51.1/M51.2 SLBMs. The French submarines, like their Western counterparts, are deployed as continuous at-sea deterrents. France's air-launched nuclear strike capability consists of two Rafale BF3 squadrons comprising around 40 fighter aircraft operated by the Strategic Air Forces (FAS) and about 10 Rafale MF3 aircraft operated by the Naval Nuclear Aviation Force (FANu) on board the *Charles de Gaulle*, the sole French aircraft carrier.[87] The strike force operated by FANu is not permanently deployed on the aircraft carrier but can be instantly deployed based on the president's assessment. The aircraft are armed with improved medium-range air-to-ground cruise missiles (ASMPA) equipped with airborne nuclear warheads (TNA).

France is developing new ballistic missile submarines called the SNL3-3G, which will replace the existing Triomphant class beginning in 2035.[88] To improve the range, accuracy, and penetration capability of the M51 SLBMs, France is developing the third iteration of the missiles, namely, the M51.3, which are scheduled to be operational in 2025.[89] Also, France is developing a stealthier and extended range replacement of the deployed ASMPA missiles known as the ASN4G to be deployed by 2035.[90]

Historically, France has sought nuclear and security policy, independent of the United States and other NATO allies. The French nuclear strategy has remained unchanged since the last defence and security White Paper was released in 2013. It is not likely that France will drastically alter its nuclear strategy soon, although it might exhibit a propensity to adapt to the emerging geopolitical and strategic environment.

## Conclusion

Three decades after the end of the Cold War, thousands of American and Russian nuclear missiles remain on high alert. Being concerned

87. Kristensen and Korda, n. 78, pp. 51-55.
88. Ibid., p. 53.
89. Ibid., p. 52.
90. Tertrais, n. 76, p. 60.

about the American missile defences and developing long-range conventional strike capabilities, China plans to expand its land-based nuclear forces. States seeking to remain relevant to the great power competition, like the United Kingdom, are looking forward to enhancing their nuclear potential. Above all, the race over hypersonic missiles and shrouded uncertainty over their nuclear capabilities amplify the risks of inadvertent escalation.

While the extension of the New Strategic Arms Reduction Treaty (START) between the US and Russia till 2026 was a positive development, the road of future arms control appears to be full of hurdles. In June 2021, the United States and the Russian Federation reaffirmed the principle that a "nuclear war cannot be won and must never be fought" and launched an "integrated bilateral strategic stability dialogue."[91] Two rounds of the strategic stability dialogue have already taken place in Geneva, Switzerland. Given that the issues that need to be discussed are extensive, many more such dialogues will be needed. Meanwhile, the US-China dynamics appears to be taking an ugly turn too. The Chinese reluctance to enter any arms control negotiations is a worrying sign about the future of nuclear arms control. Conclusively, the analysis of nuclear doctrines and vertical proliferation trends in this chapter points towards the increasing reliance of the great powers on nuclear weapons and the growing significance of nuclear weapons for the emerging great power competition.

91. "US-Russia Presidential Joint Statement on Strategic Stability", The White House, June 16, 2021. https://www.whitehouse.gov/briefing-room/statements-releases/2021/06/16/u-s-russia-presidential-joint-statement-on-strategic-stability/. Accessed on November 24, 2021.

# 6. The State of the NPT: Challenges Ahead

*Kanica Rakhra*

## Introduction

Any global debate on nuclear issues begins with the most widely followed and acknowledged treaty, the Treaty on Non Proliferation of Nuclear Weapons (NPT). Today, it has membership of more than 190 states and is the most universal treaty. The only holdouts—India, Israel and Pakistan—are possessors of nuclear weapons themselves and cannot be accommodated into the treaty.

The multilateral treaty encapsulated a "grand bargain" between the five Nuclear Weapon States (NWS) and the Non-Nuclear Weapon States (NNWS) wherein[1]

- states without nuclear weapons will not acquire them;
- states with nuclear weapons will pursue disarmament; and
- all states can access nuclear technology for peaceful purposes, under safeguards.

The three pillars, on which the treaty was based, have received unequal treatment from the member states, especially the 5 NWS. While the NWS have been stringent in ensuring that states that are party to the treaty do not acquire nuclear weapons and, in turn, have granted access to nuclear technology for peaceful purposes generously, the pursuance

1. Nuclear Threat Initiative, "NPT: Treaty on the Non-Proliferation of Nuclear Weapons (NPT)". https://www.nti.org/education-center/treaties-and-regimes/treaty-on-the-non-proliferation-of-nuclear-weapons/. Accessed on December 20, 2021.

of nuclear disarmament by the NWS has been limited, at best. Except China, the other NWS or P5 have, over the years, brought transparency into their nuclear weapon numbers and have also reduced, by a significant margin, their nuclear warheads and missiles. However, apart from making changes in the numbers, they have not truly engaged with the idea of universal nuclear disarmament.

The treaty has not been without its challenges. The USA, UK and Russia were the initial proponents of the treaty, but France and China, the other two NWS did not sign the treaty for a long time. It was only by 1995 that the treaty became near-universal with only three outlier states: India, Israel and Pakistan. Later, North Korea left the NPT. Whether it is South Asia or the Middle East, there have been multiple initiatives to bring these three states into the NPT's fold, but none has achieved success in making the near-universal NPT, universal.

Over the years, the NPT has managed to meander around the challenges and remain intact: whether it was the indefinite extension of the treaty in 1995 or the case of the Democratic People's Republic of Korea (DPRK) withdrawing from the treaty or Iran's skirting of International Atomic Energy Agency (IAEA) inspections.

However, the emergence of the Treaty for the Prohibition of Nuclear Weapons (TPNW) has caused a major stir. While the NWS are extremely wary of the treaty and what it represents, it is significant that most signatories of the TPNW are also non-nuclear member states in the NPT. They are alarmed at the ongoing modernisation of the nuclear weapon capabilities of the NWS and, more recently, of initiatives such as the Australia-UK-US (AUKUS) trilateral cooperation that will provide for a non-nuclear weapon state like Australia to acquire nuclear-powered submarines.

Will the developments of the current decade put a nail in the coffin for the NPT? Or can the treaty be reimagined. What are some of the challenges that plague the NPT? Could the growing insecurity and fears of the NWS and NNWS still be addressed within the framework of the treaty or would alternate treaties have to be crafted? This chapter examines some of the recent developments to analyse their impact on the NPT. It would seek to define the challenges ahead for the state parties in the coming decade.

### Nuclear Weapon States

The journey of each of the five nuclear weapon states—as designated by the treaty—has been unique and collaborative at the same time. As stated earlier, China and France joined the NPT at a later stage (1992) but the US, UK and USSR were keen supporters and promoters of the treaty from its initial stages. However, after the 1995 NPT Review Conference (RevCon) which made the treaty permanent, all five states have been committed to the cause of non-proliferation and disarmament and have taken various initiatives to further this cause.

Traditionally, the P5 have mostly put together a joint front at the NPT RevCons, in order to collectively ward off the criticism of the NNWS on their meagre efforts towards nuclear disarmament.[2] A similar attempt was recently evident when despite the postponement of the NPT RevCon 2020 to some time later in 2022, the P5 decided to release a joint statement[3] nonetheless. The statement reaffirms the one that had been made in 1989 by Presidents Reagan and Gorbachev: that a nuclear war cannot be won and must not be fought.

However, the changing geopolitical environment has left the P5 grappling with far more than they may have imagined—from nuclear modernisation to a renewed focus towards influx of new technologies that have the potential to disrupt traditional concepts of nuclear deterrence. In the light of all these developments, the joint statement by the P5 gives

2. For example, see UN-NPT, "Statement by The People's Republic of China, France, The Russian Federation, The United Kingdom of Great Britain and Northern Ireland, and The United States of America to the 2015 Treaty on the Non-Proliferation of Nuclear Weapons Review Conference". https://www.un.org/en/conf/npt/2015/statements/pdf/P5_en.pdf. Accessed on January 5, 2022.
3. United States, Department of State, "Joint Statement of the Leaders of the Five Nuclear-Weapon States on Preventing Nuclear War and Avoiding Arms Races", The White House, January 3, 2022. https://www.whitehouse.gov/briefing-room/statements-releases/2022/01/03/p5-statement-on-preventing-nuclear-war-and-avoiding-arms-races/. Accessed on January 5, 2022.

barely enough promise to count as a positive step in the right direction. While the statement has been welcomed by the international community and strategic thinkers around the world, its actual impact in terms of actions by the nuclear weapon states, and its impact on the TPNW, remains to be seen.

Another aspect that is likely to create a wedge amongst the P5 is the recent Australia, UK and US nuclear submarine deal or AUKUS. The deal allows for a non-nuclear weapon state (Australia) to receive technology and fissile material for a nuclear-powered submarine. It has raised many alarm bells not just in the international community[4] but also within the official positions of China and France.[5] While Russia has been careful[6] in its reactions to the alliance, Chinese officials have been regularly bringing up discussions on the subject in UN meetings.

The third aspect that the P5 need to work towards is that of finding ways to establish some sort of strategic stability, and reducing nuclear risks. While strategic stability talks between the Russian Federation and the USA have at least started, the North Atlantic Treaty Organisation

---

4. James M. Acton, "Why the AUKUS Submarine Deal is Bad for Nonproliferation—And What to Do About It" , Carnegie Endowment for International Peace, September 21, 2021. https://carnegieendowment.org/2021/09/21/why-aukus-submarine-deal-is-bad-for-nonproliferation-and-what-to-do-about-it-pub-85399; Safwan Ali, "AUKUS Deal: A Demoralizing Factor for NPT", *Modern Diplomacy,* November 16, 2021. https://moderndiplomacy.eu/2021/11/16/aukus-deal-a-demoralizing-factor-for-npt/; and John Carlson, "AUKUS Nuclear-Powered Submarine Deal—Non-Proliferation Aspects", *Asia Pacific Leadership Network (APLN)*, September 17, 2021. https://www.apln.network/projects/aukus/aukus-nuclear-powered-submarine-deal-non-proliferation-aspects. Accessed on December 20, 2021.
5. Zahra Mirzafarjouyan, "AUKUS in Breach of NPT, Damaging Regional Security: EU MP", *Tehran Times,* October 2, 2021. https://www.tehrantimes.com/news/465676/AUKUS-in-breach-of-NPT-damaging-regional-security-EU-MP. Accessed on December 20, 2021.
6. PTI, "After AUKUS, Russia Sees a Potential Threat - and an Opportunity to Market its Own Submarines", *Economic Times,* September 25, 2021. https://economictimes.indiatimes.com/news/defence/after-aukus-russia-sees-a-potential-threat-and-an-opportunity-to-market-its-own-submarines/articleshow/86505017.cms. Accessed on December 20, 2021.

(NATO)-Russia dialogue is still at a stalemate as the latter is keen to include the UK's nuclear arsenal in the discussions, something that the UK does not wish to engage in. The prospects of NATO-Russia engagment have been further marred by the Russia-Ukraine conflict. China refuses to engage in any bilateral or trilateral talks and would want the discussion to be held amongst the P5. However, France and the UK, that have been relatively transparent in their nuclear developments, do not believe they should be a part of any strategic stability talks. Additionally, both states have embarked on nuclear modernisation of their forces and the UK has also begun to increase its nuclear arsenal.

The P5 are currently both converging and diverging on different aspects of their NPT goals, which makes it difficult for them to not only engage within the NWS but more so with the NNWS. However, the P5 are equally invested in keeping the NPT as the main pillar of the global nuclear architecture. Without the NPT, many legal frameworks of the Nuclear Suppliers' Group (NSG), and the International Atomic Energy Agency (IAEA) would be found wanting and it may lead to further problems as a universal treaty on the lines of the NPT may be difficult to evolve in the current global scenario.

## TPNW

The Treaty on the Prohibition of Nuclear Weapons (TPNW) has come as a new development in the global nuclear architecture. While the ardent champion of disarmament, former Sri Lankan Ambassador Dhanapala[7] calls it "a ray of hope", the more practical nuclear analyst Daryl Kimball believes that all the nine nuclear weapon states "would be better served to publicly recognize its arrival as a good faith effort by the majority of states to eliminate the nuclear danger and build up the legal framework for the elimination of nuclear weapons".

The consistent delay on the part of the P5 of the NPT to work towards their goal of disarmament, as defined in the treaty, has led numerous states to come together and formulate a new treaty in the

7. Jayantha Dhanapala, "Finally, Nuclear Weapons are Outlawed", *UNIDIR*. https://unidir.org/node/5720. Accessed on December 25, 2021.

form of the TPNW. The TPNW member states overlap with the NPT member states and representatives from the former always insist that the TPNW is more of an extension of the NPT. Kimball praises that treaty for including aspects such as "victim assistance and environmental remediation to those affected by nuclear weapons use and testing".[8] Hajnoczi analyses both the treaties, and states that the TPNW, "explicitly and structurally fits into the framework created by the NPT and constitutes a necessary measure for the implementation of its Article VI".[9] He also highlights how the TPNW "uses the terminology of the NPT and other established disarmament treaties regarding weapons of mass destruction".

Erasto echoes this thought, stating that INFCIRC/153 (International Atomic Energy Agency Information Circular/153) (Corrected) in the TPNW can nevertheless be seen as an improvement from the NPT Article III, which does not specify any particular safeguards standard.[10] Furthermore, in their paper, Nystuen, Egeland and Hugo clarify in detail how the TPNW does not supersede the NPT and in fact, plugs the gaps in the NPT.[11] The TPNW then, is akin to the NSG or the IAEA in that it supports implementation of the NPT, rather than act as a challenger. This was reiterated by Ambassador Gustavo Zlauvinen during his talk at the Oslo Nuclear Project. He pointed out that "there is a constellation of treaties, instruments and agreements that make up

8. Daryl Kimball, "The Nuclear Ban Treaty: A Much-Needed Wake-Up Call", *Arms Control Today,* November 2020. https://www.armscontrol.org/act/2020-11/focus/nuclear-ban-treaty-much-needed-wake-up-call. Accessed on December 25, 2021.
9. Thomas Hajnoczi, "The Relationship between the NPT and the TPNW", *Journal for Peace and Nuclear Disarmament,* March 3, 2020. https://www.tandfonline.com/doi/full/10.1080/25751654.2020.1738815. Accessed on December 25, 2021.
10. Tytti Erasto, "The NPT and the TPNW: Compatible or Conflicting Nuclear Weapons Treaties?" *SIPRI Blog,* March 6, 2019. https://www.sipri.org/commentary/blog/2019/npt-and-tpnw-compatible-or-conflicting-nuclear-weapons-treaties. Accessed on December 25, 2021.
11. Gro Nystuen, Torbjrn Klv Egeland and Hugo Graff, "The TPNW: Setting the Record Straight", Norwegian Academy for International Law, October 2018. http://intlaw.no/wp-content/uploads/2018/10/TPNW-Setting-the-record-straight-Oct-2018-WEB.pdf. Accessed on December 25, 2021.

the nuclear disarmament and non-proliferation regime, but the NPT remains its beating heart".[12]

There are other arguments however, that question the co-binding of the NPT and TPNW. For example, Mueller and Wunderlich argue that the TPNW should be seen as an "act of self-empowerment: nuclear have-nots produced an international disarmament treaty without the involvement of the nuclear weapon states or their allies".[13]

However, all these factors and arguments have not found favour with the five nuclear weapon states. They are still very wary of discussions leading up to the TPNW. The US has engaged with its nuclear umbrella states to ensure minimal participation.

The *Nuclear Weapons Ban Monitor*, published by the Norwegian People's Aid, monitors progress on compliance with, and support for, the TPNW by producing a country-by-country analysis and an overview of the status of nuclear weapons today. The *Nuclear Weapons Ban Monitor* categorises 135 countries as TPNW supporters and 31 countries as nuclear weapon-complicit states, meaning they have endorsed, or acquiesced to, the possession and potential use of nuclear weapons on their behalf.[14]

However, owing to the continuous work being done by the TPNW member states along with the International Campaign Against Nuclear Weapons (ICAN), more NATO members were becoming keen to participate, if not as full members, then at least as observers. Along with the 58 state parties of the TPNW, some NATO members that may be participating in the upcoming TPNW conference in March 2022 are likely to include Germany, Sweden, Norway and, possibly,

12. Gustavo Zlauvinen, "Challenges to the NPT", *Oslo Nuclear Forum 2020*, September 16, 2020. https://www.un.org/sites/un2.un.org/files/olso_nuclear_forum_gz_speech_-_16_september_2020.pdf. Accessed on December 25, 2021.
13. Harald Müller, and Carmen Wunderlich, "Nuclear Disarmament Without the Nuclear-Weapon States: The Nuclear Weapon Ban Treaty", *Daedalus*, 2020. https://www.amacad.org/sites/default/files/publication/downloads/Daedalus_Sp20_11_Muller%20%26%20Wunderlich.pdf. Accessed on December 25, 2021.
14. ICAN, "Two-Thirds of the World's Countries Support TPNW", International Campaign to Abolish Nuclear Weapons. https://www.icanw.org/two_thirds_of_the_world. Accessed on December 25, 2021.

Canada. It remains to be seen whether the ongoing Russia-Ukraine conflict will influence their decision.

## NATO and Nuclear Umbrella States

Recent research by analyst Leopoldo Nuti indicates how NATO member states, as nuclear umbrella states, were part of the negotiations leading up to the NPT and that the language of the NPT was designed specifically to take into consideration the security concerns of NATO's European partners.[15] Although NATO is neither a party nor a signatory to the NPT, and the alliance does not possess nuclear weapons, its member states are legally bound by international arrangements. The text of the NPT, however, was discussed within NATO's Nuclear Planning Group (NPG) and was designed to ensure that the NPG could function independently, without being questioned by the then USSR, and later the Russian Federation. Another aspect that was catered for in the text of the NPT was the ability of the United States of America to deploy its nuclear weapons in allied territory as long as they remained under US control, with no physical nuclear sharing.

This research puts the TPNW at loggerheads with the NPT as the former treaty focusses solely on elimination of nuclear weapons and does not engage with aspects of deterrence and security. This was also raised by Jens Stoltenberg, in November 2020, at NATO's Annual Conference on Weapons of Mass Destruction, Arms Control, Disarmament and Non-Proliferation where he stated:[16]

> I know that there are those that look at the Treaty on the Prohibition of Nuclear Weapons—or the Ban Treaty—as an alternative solution to eliminate all nuclear weapons. At first glance, it seems attractive. But the reality is that it will not work. The Ban Treaty has no mechanism

15. Leopoldo Nuti, "NATO's Role in Nuclear Non-Proliferation and Arms Control: A (Critical) History", Instituto Affari Internazionali, January 2021. https://www.iai.it/sites/default/files/iai2103.pdf. Accessed on December 20, 2021.
16. NATO, "Speech by NATO Secretary General Jens Stoltenberg at the 16th Annual NATO Conference on Weapons of Mass Destruction, Arms Control, Disarmament and Non-Proliferation", 2020.

> to ensure the balanced reduction of weapons. And no mechanism for verification. Moreover, it has not been signed by any state that possesses nuclear weapons. Simply giving up our deterrent without any guarantees that others will do the same is a dangerous option.

Additionally, the December 2020 North Atlantic Council (NAC) statement recalls that:[17]

> NATO is a defensive alliance. The fundamental purpose of NATO's nuclear capability is to preserve peace, prevent coercion, and deter aggression. A world where the states that challenge the international rules-based order have nuclear weapons, but NATO does not, is not a safer world. [...] We call on our partners and all other countries to reflect realistically on the Ban Treaty's impact on international peace and security, including on the NPT, and join us in working to improve collective security through tangible and verifiable measures that can reduce strategic risks and enable real progress on nuclear disarmament.

Even though TPNW member states have signified at length that the treaty is not antithetical to the NPT and that it should be seen as an extension of the NPT, the fear within NATO of not addressing the security concerns as was done during the NPT negotiations, still remains. Caughley and Afina[18] highlight that at the beginning of the TPNW negotiations, the Netherlands also made it "clear that any prohibition must be compatible with the NATO principles of the Deterrence and

17. NATO, "North Atlantic Council Statement as the Treaty on the Prohibition of Nuclear Weapons Enters Into Force", 2020.
18. Tim Caughley, with Yasmin Afina, "NATO and the Frameworks of Nuclear Non-proliferation and Disarmament Challenges for the 10th NPT Review Conference", *Chatham House,* May 2020. https://www.chathamhouse.org/sites/default/files/2020-05-29-nato-npt-frameworks-caughley-afina-2.pdf. Accessed on December 25, 2021. Also see The Netherlands (2017), "United Nations Conference to Negotiate a Legally-Binding Instrument to Prohibit Nuclear Weapons, Leading Towards their Total Elimination: Agenda Item 8 (b): Statement by The Netherlands", March 28, 2017. http://statements.unmeetings.org/ media2/14683480/netherlands.pdf

Defence Posture Review, including the notion that NATO will remain a nuclear alliance as long as nuclear weapons exist". More recently, the German foreign minister stated, "A world without nuclear weapons is a safer world. We want to make a contribution to this and—in consultation with partners—participate as an observer at the TPNW. This does not weaken our commitment to NATO and nuclear sharing."[19]

Hach believes that nuclear umbrella states such as Germany, Japan, Australia and Canada, while being important members of the NPT, also enjoy a certain degree of closeness with three of the officially designated nuclear weapon states (the USA, UK and France).[20] Hach further points out that they may be able to use their closeness to the nuclear weapon states to pursue each of their personal state goals of disarmament by engaging with the TPNW. Many have already begun the process, with Swedish Foreign Minister Ann Linde giving historical references regarding bridges between the NWS and NNWS.[21] But its success and/ or failure remains to be seen.

## Other Challenges

Apart from the challenges that have come to be a part of the NPT owing to the cleavages between the NWS and the NNWS and the arrival of the TPNW, there are a few other stresses too that the treaty faces.

### *Withdrawal Clause*

Article X of the NPT provides a "right" to withdraw from the treaty if the withdrawing party "decides that extraordinary events, related to the

19. Hieko Mass, "Statement on German Foreign Office", Official Twitter account https://twitter.com/GermanyDiplo Tweet on December 14, 2021.
20. Sascha Hach, "The Art of Engineering at the NPT Review Conference. How Germany and Other Umbrella States Can Build Bridges", *PRIF Blog*, December 9, 2021. https://blog.prif.org/2021/12/09/the-art-of-engineering-at-the-npt-review-conference-how-germany-and-other-umbrella-states-can-build-bridges/. Accessed on December 20, 2021.
21 Ann Linde, "Can Disarmament Be Revived? An Interview With Swedish Foreign Minister Ann Linde", Arms Control Association, May 2021. https://www.armscontrol.org/act/2021-05/features/disarmament-revived-interview-swedish-foreign-minister-ann-linde. Accessed as on December 25, 2021.

subject matter of this [t]reaty, have jeopardized the supreme interests of its country."[22] It also requires that a withdrawing state-party give three months' notice. Additionally,

- it must give notice of withdrawal to all parties to the NPT;
- it must give notice of withdrawal to the United Nations Security Council;
- it must provide a statement of the extraordinary events which the state party considers to have jeopardised its supreme interests; and
- the withdrawal notice must be provided 90 days in advance of the effective withdrawal date.[23]

The withdrawal clause of the NPT has been contentious and was the cause of many debates prior to the treaty coming into being; during the Eighteen-Nation Disarmament Committee (ENDC) talks as well. But given that there were many states that needed to be convinced to join the NPT, the withdrawal clause was an important consideration that could not be removed from the equation.

However, the use of the withdrawal clause by the DPRK first in 1993, and then in 2003, raised a different set of questions for the future of non-proliferation and the NPT. The DPRK renewed its 1993 notice of withdrawal from the NPT, a notice that had been suspended a decade earlier during negotiations with the United States, in 2003. This left the UN Security Council with only a single day before North Korea would become the first country to withdraw from the NPT.[24] The state exploited the withdrawal clause and gave no space for negotiations or discussions,

22. UN Documents, "The Nuclear Non-Proliferation Treaty". https://www.un.org/disarmament/wmd/nuclear/npt/text/. Accessed on December 25, 2021.
23. Jenny Neilson and John Simpson, "The NPT Withdrawal Clause and its Negotiating History", Mountbaiten Centre for International Studies, July 2004. https://eprints.soton.ac.uk/39771/1/withdrawal_clause_NPT_nielsen%2526simpson_2004.pdf. Accessed on December 20, 2021.
24. George Bunn and John B. Rhinelander, "NPT Withdrawal: Time for the Security Council to Step in", Arms Control Association, https://www.armscontrol.org/act/2005-05/features/npt-withdrawal-time-security-council-step. Accessed on December 20, 2021.

especially during the 2003 decision. Many analysts argue that the DPRK is still a part of the NPT as its withdrawal was hasty and did not follow the correct procedure.

The DPRK's withdrawal from the NPT, the difficulties that the international community has had in addressing that withdrawal, and the failure to find a lasting resolution to the disputes over the North Korean programme, constitute significant challenges to the NPT. However, a bigger challenge for the NPT is to ensure that no country follows in these footsteps.

Debates about whether Iran may choose such a route have surfaced every now and then. In the past, these concerns were often dismissed by highlighting that Iran's regional ambitions differed from those of North Korea. The latter was believed to be content in its region and did not harbour aspirations of regional and/or international domination. However, the fallout from the Joint Comprehensive Plan of Action (JCPOA) has changed internal Iranian thinking on the issue significantly.

Iran has, in the recent past, threatened to withdraw from the NPT twice (2018[25] and 2019[26]). These statements by the foreign minister signify a shift in thinking within the Iranian polity which does not bode well for the international non-proliferation regime. Additionally, the recent statement by Ali Larjiani[27] adds to the growing consensus within not just hardline but also moderate Iranian voices.

---

25. Bozorgmehr Sharafedin, "Iran Warns Trump it Might Withdraw from Non-Proliferation Treaty", Reuters, April 24, 2018. https://www.reuters.com/article/us-iran-nuclear-rouhani/iran-warns-trump-it-might-withdraw-from-non-proliferation-treaty-idUSKBN1HV0MY. Accessed on December 20, 2021.
26. Reuters, "Iran says Leaving Nuclear NPT One of Many Choices After US Sanctions Move", *Japan Times,* April 28, 2019. https://www.japantimes.co.jp/news/2019/04/28/world/politics-diplomacy-world/iran-says-leaving-nuclear-npt-one-many-choices-u-s-sanctions-move/#.%20XSHebBNKi9Y. Accessed on December 20, 2021.
27. Borzou Daragahi, "Iran Tensions: Tehran Warns it will Leave Global Nuclear Weapons Treaty if Referred to UN", *The Independent,* January 20, 2020. https://www.independent.co.uk/news/world/middle-east/iran-tensions-nuclear-weapon-treaty-missile-non-proliferation-a9291696.html. Accessed on December 20, 2021.

Saudi Arabia has already stated that it will follow Iran in pursuing the nuclear bomb.[28] This, in turn, may lead to a domino effect in the region with Egypt and Turkey seeking their own nuclear programmes.

Thus, the DPRK's withdrawal may have caused debates about the withdrawal clause, but it will not have an impact on other member states of the NPT if they decide to withdraw. The Iranian ambitions, if proved successful, would have serious ramifications for the region and the global non-proliferation order.

### *Universality*

As the NPT completes 52 years even while being ridden with so many problems, it is important to acknowledge the reason for its success all these years. With 191 member states, the NPT is the only near-universal treaty dealing with issues of non-proliferation and disarmament. But has its universality brought structure to, or become a roadblock in, ushering progress in the international nuclear regime?

The global nuclear order is experiencing not just horizontal but also vertical proliferation. All the P5 states are currently developing their nuclear arsenals and the focus within the international strategic community has shifted to the development of a secure second strike capability for not just the USA and Russian Federation, but also for the remaining state parties of the P5.[29] While the NWS such as the UK are increasing their nuclear arsenals, the Black Sea states, that are part of NATO, as well as NATO aspirants such as Ukraine, are calling for added

---

28. Reuters, "Saudi Crown Prince says will Develop Nuclear Bomb if Iran Does", Reuters, March 15, 2018. https://www.reuters.com/article/us-saudi-iran-nuclear-idUSKCN1GR1MN. Accessed on December 20, 2021.
29. Rose Gottemoeller, "The Standstill Conundrum: The Advent of Second-Strike Vulnerability and Options to Address It", *Texas National Security Review,* August 2021. https://tnsr.org/2021/10/the-standstill-conundrum-the-advent-of-second-strike-vulnerability-and-options-to-address-it/. Accessed on December 25, 2021.

security.[30] The European Union (EU) states too, are keen to engage on matters of security[31] with many states releasing their own Indo-Pacific strategies.

Even the discussions at the Track 1.5 level do not look promising. Tired with the quagmire of questions on security and the dream of elusive disarmament, the discussions have shifted to racism within the strategic community.[32]

With the current developments in the nuclear order, engaging the three outlier nuclear weapon states has become a possibly defunct topic. The nuclear trajectory of the three states, India, Israel and Pakistan, has been different from the nuclear journey of the P5 states. The P5 states have the added responsibility of reaching out to the NNWS of the NPT and working towards the goal of disarmament. It seems unlikely that the three outlier states would want to become a party to the treaty and take

30. Abraham Mahshaie, "Black Sea NATO Allies Call for Added Security Amid Russian Buildup", *Air Force Magazine,* December 17, 2021. https://www.airforcemag.com/black-sea-nato-allies-call-for-added-security-amid-russian-buildup/?utm_source=Sailthru&utm_medium=email&utm_campaign=12.20.2021&utm_term=Editorial%20-%20Early%20Bird%20Brief; Michael Sirwitz, "NATO Signals Support for Ukraine in Face of Threat From Russia", *New York Times,* December 16, 2021. https://www.nytimes.com/2021/12/16/world/europe/ukraine-nato-russia.html?utm_source=Sailthru&utm_medium=email&utm_campaign=EBB%2012.17.2021&utm_term=Editorial%20-%20Early%20Bird%20Brief; Reuters, "EU, Germany, UK Send New Warnings to Russia over Ukraine", *Al Jazeera,* December 15, 2021. https://www.aljazeera.com/news/2021/12/15/eu-germany-uk-send-new-warnings-to-russia-over-ukraine?utm_source=Sailthru&utm_medium=email&utm_campaign=EBB%2012.16.2021&utm_term=Editorial%20-%20Early%20Bird%20Brief; and Reuters, "Germany says Russia Will Face Massive Consequences if it Invades Ukraine", December 14, 2021. https://www.reuters.com/world/europe/germany-says-russia-will-face-massive-consequences-if-it-invades-ukraine-2021-12-14/?utm_source=Sailthru&utm_medium=email&utm_campaign=EBB%2012.15.2021&utm_term=Editorial%20-%20Early%20Bird%20Brief. Accessed on December 25, 2021.
31. "The Political Impact of the European Military Presence in the Indo-Pacific", Japan Institute for International Affairs, November 8, 2021. https://www.jiia.or.jp/eventreport/2021-11-10-teip.html. Accessed on December 25, 2021.
32. "Nuclear Policy in the Biden Administration and Beyond", *Ploughshares Fund,* October 21, 2021. https://vimeo.com/638813508. Accessed on December 25, 2021.

on the added responsibility of outreach towards the NNWS. It seems equally unlikely that the P5 would be interested in bringing more nuclear weapon states into the fold of the NPT and risk upsetting the already shaken foundations of the treaty.

## Conclusion

The global nuclear order is in a state of flux; the P5 are developing nuclear weapons, and the NWS and NNWS are unable to communicate with each other, while new treaties such as the TPNW are bringing new dimensions to the nuclear order.

The NPT Review Conference (RevCon), which is held every five years, releases a final document. With the 2000 and 2010 documents as the background, the wait for the 2020 RevCon was at its peak when it got postponed due to the pandemic. However, two years was a long enough wait for the P5 to come together and release a statement on the futility of a nuclear war.

Future RevCons will need to concentrate on building bridges between the NWS and NNWS, among the NWS, as well as among the NNWS. This is the only way of retaining the success of non-proliferation, which cannot be done without credible movement towards disarmament. The challenge of the unequal pillars of the NPT will have to be addressed at some point in the near future if the NPT is to retain its status as the bedrock of the international non-proliferation architecture.

# 7. The Iran Nuclear Imbroglio: End in Sight?

*Sheel Kant Sharma*

There is an element of waiting with bated breath about the return to the Joint Comprehensive Plan of Action (JCPOA) by the USA and Iran. More than one year since President Biden's inauguration and his promise to return to the nuclear deal, several rounds of talks in Vienna have not yet yielded a positive result. Even though for more than a quarter century, the nuclear issues with Iran have figured on the front burner, with implicit threats of military confrontation, the time seems to be running out for diplomacy as Iran's hardline government under President Raeesi appears less inclined to follow former President Rouhani in negotiating a return to the 2015 nuclear accords.[1] Washington too, despite President Biden's promises about retrieving the JCPOA, seems susceptible to multiple pressures, both domestic and geopolitical. In the event, an optimistic Joint Statement issued by the US and its allies in Rome on October 31, 2021, at the margins of the G20 Summit, had served to dispel uncertainty. The Joint Statement contained President Biden's pledge that the US will not leave the deal if Iran too complies with it.[2] Iran's official announcement of the date and statement from the US State Department spokesman about the approach to these talks instilled optimism.[3]

1. Joint Comprehensive Plan Of Action. Its text is given in full in the Annex to UN Security Council Resolution 2231. https://undocs.org/S/RES/2231(2015). The resolution comprises all the elements of the accord in its operative part while the JCPOA as an integral part, is attached to it.
2. https://www.theguardian.com/world/2021/oct/31/us-iran-nuclear-deal-tehran-joe-biden-pledge
3. https://edition.cnn.com/2021/11/03/politics/iran-nuclear-talks-restart-date/index.html?mkt_tok=MDk1LVBQVi04MTMAAAGAi56EQmlo

There was overall confluence of mutual interests in 2014-15 that had led the US and Iran to forge the nuclear accord, with support forthcoming from the US allies, as well as Russia and China. That confluence became more and more tenuous after President Trump withdrew from the deal. Numerous factors have cropped up again. Trump's scrapping the JCPOA made Iran more and more sceptical and hardened its stance. The sharp downturn in great power relations has been of no help. That downturn obscured, among other things, the boon to the global non-proliferation order that the JCPOA's success would have granted.

At the same time, there have been straws in the wind. For example, the US and Russia, since agreeing to the extension of the New Strategic Arms Reduction Treaty (START) Treaty in January 2021, had registered several high level contacts, thus, relieving some part of the bilateral tensions, particularly in the nuclear context. This seemed to have helped the atmospherics for negotiations with Iran. For instance, even though Russian President Putin avoided the Rome G-20 Summit, Russia did welcome the US-E3 pledge. A senior US official was quoted by the *New York Times* as saying that "Russia had been closely aligned with the United States on restoring the Iran nuclear deal".[4] The European Union (EU) which is coordinating the Vienna talks has officially confirmed consent from all the participants.[5] US-China relations have been on a downward slope in general while Iran had forged long-term partnerships with China, as well as Russia. Even so, the expectation was that the success of the diplomacy of the P5+ Germany/EU with Iran might serve their interests better.

The Russian invasion of Ukraine, the ferocity of the war unleashed, absence of any sign of its end in sight and the sinister tenacity of the Russian offensive have dealt a body blow to the global system where diplomacy offered hope against pitfalls of geopolitics. This might bedevil the fate of the dialogue with Iran. Many assumptions underlying the entire diplomatic process of grappling with the Iranian imbroglio have come under severe strain. Whether sanctions can compel, what

---

1Qc-CpIF6XjJKoQcejAfbJTcquhXhwAKRDCvh3hy7QdKBVnwQBe_Ec3j3DGseOmeyhSzC4C12hPuHJtWNaWKsyBmveikP7_tcEw

4. https://www.nytimes.com/2021/10/31/world/europe/biden-putin-russia-united-states.html?action=click&module=RelatedLinks&pgtype=Article

5. n. 3.

guarantees security for those faithful to nonproliferation order (if there is one!), what would be for Iran a trade off worth, and whether the P5 can stick together all the way – these are questions that will loom large.

## Understanding the Full Perspective

It would be useful to gain a fuller perspective by revisiting the many phases through which Iran and the USA have been over the past three decades as also the evolution of the nuclear programme of Iran.

### *Phase 1*

During the decade following the revelation of Iraq's clandestine nuclear programme from 1990-91 to 2002-03, Iran began to loom large as a potential challenge to the global non-proliferation order. It was during this phase that the US had placed non-proliferation diplomacy on an overdrive. President George H. Bush even chaired a meeting of the UN Security Council on January 31, 1992, where a resolution was adopted affirming nuclear proliferation as a threat to international peace and security. Moreover, the gross violations of the Treaty on Non-Proliferation of Nuclear Weapons (NPT) by Iraqi President Saddam Hussein paved the way to bring regional politics into the nuclear proliferation issues, transcending the legal framework of the NPT. It was, thus, getting harder to draw a line between extraneous political manoeuvring and NPT implementation. It was even considered by non-proliferation enthusiasts that the UN Security Council Resolution 687 (which had laid down a comprehensive disarmament plan of action, albeit limited solely to Iraq) should also be applied to other proliferation cases. These trends had a considerable adverse impact on Iran during the 1990s.

The nuclear and missiles proliferation allegations against Iran emerged almost simultaneously with the detection of Saddam Hussein's clandestine nuclear weapons pursuit in 1990. The decade of the Nineties witnessed a mounting political campaign claiming that Iran was secretly working for the bomb. Riding on the tide, as it were, Benjamin Netanyahu, then Israeli envoy in Washington declared in 1995 that Iran was just a few years from acquiring a nuclear weapon. Iran's push for nuclear power reactors—which actually dated back to the 1970s under the Shah of Iran—fuelled the perception that it was as a foil for work on a nuclear weapon.

It was argued that since Iran had amongst the largest hydrocarbon reserves globally, its quest for nuclear energy strained credulity. This line had weight in the decade after the Chernobyl accident, on the one hand, and Saddam Hussein's clandestine pursuit, on the other. Nuclear power's stock had sunk low as an unsafe energy source and for being a route for the bomb. Iran, nonetheless, proclaimed its inalienable right as a party to the NPT to work on a nuclear fuel cycle and to build nuclear power reactors. This somehow provided grist to the right wing campaign in the US and Israel, and the West in general. The evidence about Iraqi nuclear enrichment work hidden from the International Atomic Energy Agency (IAEA) despite Iraq being a party to the NPT was advanced as the reason to suspect others too, most notably Iran. Iran's rejection of the strengthened safeguards of the IAEA, including the 1997 Additional Protocol, made such suspicions stronger.

It is noteworthy that the entire process within the IAEA from 1993 onward to systematically strengthen its safeguards system aimed to provide early warning about undisclosed non-compliance. This, in turn, was motivated to rule out cheating along the lines of what had transpired in Iraq. Such a process had widespread support among IAEA members, particularly the NPT parties. So, Iran's non-adherence to stronger safeguards stood out, even though this could be a legally defensible stance predicated on the NPT's silence on anything additional to comprehensive safeguards as enshrined in the IAEA document International Atomic Energy Agency Information Circular/153 (INFCIRC/153). Be that as it may, no tangible proof was available until 2002 about Iran actually hiding alleged non-compliance with NPT safeguards.

### *Phase 2*

Between 2002 and 2006, credible evidence surfaced from diverse sources.[6] Questions were raised within the IAEA Board of Governors about Iran's concealment of nuclear activities with respect to lack of

6. Iranian dissidents, US intelligence and IAEA reports.

compliance with the NPT and the related safeguards agreement, and the panic that was created about the threat of US-Israeli military action. This was viewed as having grave implications, especially in the context of the then raging US War on Terror and US military action against Iraq. What followed were the most difficult years of an increasingly precarious stalemate, particularly from 2005 to 2012. Tehran had the most hardline regime of President Ahmadinejad during this phase. Iran defied the sanctions, braved mounting threats of the US/Israeli military option, and succeeded in building up its centrifuges to well beyond 18,000 to enrich uraniumn. As the US suffered serious reverses in its War on Terror, this tempered the hawks in Washington and the White House looked for other options. Still, an intense outbreak of armed conflict between Israel and the Hezbollah in 2006 lent its own dynamics. Israel attacked and destroyed in 2007 a nuclear facility allegedly being clandestinely built in Syria. The IAEA admitted at that time that it was in no position to carry out inspections to verify what would have been underway in the destroyed facility. The Obama Administration consented to the European allies talking to Iran. The US also carried out a carefully planned secret cyber attack[7] on the Iranian enrichment plant where hundreds of centrifuges were knocked out. This attack was later admitted by US commentators as Obama's option to put pressure on Iran, short of a military action. Although this was a setback, Iran was able to bring its centrifuges back into action in a year's time.

### *Phase 3*

This phase was followed by one of fervent negotiations under the shadow of the military option from 2012-15. This was through direct talks with Iran by the P5 plus Germany/EU as well as a back channel between the US and Iran.[8] These negotiations, finally, led to the

7. David E. Sanger, *The Perfect Weapon* (New York: Crown, 2018). This book provides ample details about the operation, Olympic Games, which was carried out covertly and deniably by the US National Security Agency (NSA) and Israeli Unit 8200 to blow up Iranian centrifuges by using a malware called Stuxnet.
8. William Burns, *The Back Channel* (New York: Random House, 2019).

conclusion of the JCPOA. The short period of relief and satisfaction with implementation of the JCPOA by Iran supported by the P5 plus Germany/EU ended when the US Administration under President Donald Trump disrupted the JCPOA implementation. From 2017 onward, his Administration systematically undermined, revoked and dismantled the JCPOA and associated confidence- building measures that the Obama Administration had put in place through executive orders. Trump reimposed sanctions on Iran and proceeded to tighten them. The JCPOA had to bypass the Congressional route of ratification and was put into force by an executive order because President Obama was not sure of mustering adequate majority.

It is only since January 2021 that revived diplomatic efforts by France, the UK, Germany, EU, Russia and China have been underway in Vienna (with US consent) to reach an agreement with Iran to return to the JCPOA. Six rounds had taken place by June 2021 before the Iranian elections. Iran and the US, in conjunction, have held indirect talks in the same venue. It was a general impression in June 2021 that both sides were close to agreeing to return to compliance with the JCPOA—although Iran still demanded credible assurances against a repetition of disruptions, like those wrought by Trump, under a future US Administration. President Biden appears unable to provide such guarantees.

The analysis of issues germane to each phase reveals a *leitmotif* that has emerged through the years: a *leitmotif* that nuclear proliferation was a handy ploy for the hardliners to permanently place Iran under the spotlight for unending sanctions and isolation. Equally, Iran doggedly assessed that the nuclear issue was a potent means for national consolidation and sought a hard bargain, *inter-alia*, to emerge out of the isolation faced by it. The aim in this chapter is to draw inferences from the international community's experience about grappling with the consequences of Iran's nuclear programme, and diplomatic efforts to come to terms with it as well as coping with the menacing potential of the situation getting out of control, either with further weaponisation or military conflict. It is then attempted to envisage the praxis.

The JCPOA, it is relevant to recap, had essentially confined the agreement to Iran's comprehensive commitments in the nuclear domain in return for lifting of sanctions imposed by the US, EU and UN Security Council in broad sequential steps. The UN Security Council Resolution 2231[9] unanimously and unequivocally took on board the JCPOA and granted it the seal of UN authority. The signal contribution of the JCPOA to the global non-proliferation order was in the willing cooperation of Iran in its transparent implementation, with continuous monitoring of compliance and verification certified by the IAEA. The IAEA's entire role as spelt out in the JCPOA has been way beyond the commitments specific to Iran's NPT safeguards (INFCIRC 153) and an associated Additional Protocol or indeed concerning the IAEA's statutory remit. Rather, the IAEA was also tasked for the first time to monitor activities related to weaponisation.

Iran was duly complying with these larger commitments when Trump revoked the JCPOA in May 2018 and reimposed the sanctions. As later accounts by President Obama[10] and his Deputy Secretary of State William Burns[11] [now Central Intelligence Agency (CIA) director under President Biden] in their respective memoirs have testified, the wider concerns about Iran's military and missiles and their allegedly destabilising role in the Middle East could not find a place in the 2015 final deal. Nor did Iran let such 'non-nuclear' issues encroach into the deal .

This is where the rub lay about the delicate deal and the conflicted plight of an Iran not being on good terms with its neighbours,especially Israel. The US played a lead role in trying to stabilise the situation in 2015 so as to avoid a military conflagration.

It is important to recap what the full implications of the JCPOA were at the time of its conclusion in 2015.

The nuclear accord was concluded on July 14, 2015 (between Iran and the P5 plus Germany/EU). It was followed, within days, by the passing of a consensual UN Security Council Resolution 2231. It

9. https://undocs.org/S/RES/2231(2015)
10. Barak Obama, *A Promised Land* (Viking, 2020), pp. 453-454.
11. Burns, n. 8, ch 9, "Iran and the Bomb: The Secret Talks", pp. 336-387.

was widely seen as a truly historic and landmark achievement not only for non-proliferation but also for diplomacy and multilateralism. The JCPOA was the first of its kind in the annals of non-proliferation and arms control. It was anchored on the NPT and the associated safeguards agreement of Iran. However, it was unique in the extraordinary sweep of Iran's negotiated, yet voluntary, commitments about closing the diverse pathways to nuclear weapon capability; as also drastic scaling down and shuttering of its uranium enrichment activities under the IAEA's most intensive monitoring, verification and control. The JCPOA comprised a main text and five annexes which were integral to the accord. These were even likened to the extensive arms control treaties of the Cold War years in leaving virtually nothing ambiguous.

In fact, Iran had agreed to go so much out of its way in assuming the obligations under the JCPOA that the document had an explicit disclaimer clause, possibly, to reassure the others in the Middle East, if not the wider world, that it would not be precedent-setting in any manner. However, the immense diplomatic worth of the success of the almost decade-long on-off negotiations far exceeded the detailed minutiae of the deal. The IAEA painstakingly implemented verification and confirmed Iran's compliance in January 2016. This provided at that time a further boost to the credibility and effectiveness of the IAEA beyond the Nobel Prize of a decade earlier. It added respectability to the UN Security Council, which in the Middle East in particular, had suffered from a chequered record. This nuclear deal signified much more than a silver lining at a time which was particularly bleak for arms control and non-proliferation. Its annexes raised the standards of a cooperative process of non-proliferation in one of the more difficult regions confronted with a precarious security situation.

The so-called break-out time, the time that Iran was expected to take to acquire nuclear weapons, was effectively increased to well over a year. It was assumed that such a prolongation would give adequate space and time to the US and its allies for invoking other options. Iran agreed to severe restrictions that capped the quantity and level of uranium enrichment at its plants in Natanz and Fordow, and for removing about two-thirds of its centrifuges. Its stockpile of enriched uranium was brought down to just 300 kg and the level of enrichment capped at 3.67 per cent. All enrichment was ceased except at the centrifuges at Natanz, and strict limits were placed on the number and sophistication of the operating centrifuges and on heavy water production. The JCPOA provided for continuous monitoring of centrifuges and centrifuge rotor tubes, monitoring of heavy water stock, continuous access to Natanz, monitoring of the production or acquisition of uranium ore concentrate, enhanced access on activities related to the conversion to uranium hexafluoride gas as well as any further steps on the metallurgy of enriched uranium. All this was intended to ensure that the IAEA would be able to detect a suspected violation.

Iran was also obliged to shutter its plutonium production heavy water research reactor at Arak (IR-40) and to stop work on uranium metal. As Iran's mining and milling of uranium was placed under the IAEA's continuous surveillance, this was expected to ensure that no unmonitored diversion to non-peaceful use would take place. All nuclear facilities in Iran were under safeguards and Iran was obliged to disclose design information of any new installations right at the planning stage under Code 3.1 of the enhanced 153 type safeguards. Thus, every step, from mining to fabricating uranium metal in any form, was covered by IAEA monitoring and verification, So, the likelihood of cheating by Iran to divert material and technology towards weapons use was minimised. Also, limitations were placed on Iran's long-term civilian nuclear activities [like Research and Development (R&D)] in order to further tighten its commitment to never acquire nuclear weapons . The IAEA monitored and verified compliance with these commitments in January 2016 which led, in turn, to relaxation of a tranche of sanctions on Iran. Approximately revenues from oil sales worth $ 4.2 billion were released.

This kind of comprehensive accord was hailed, among other things, for being a pointer to peacefully resolving problems in regard to other regional issues concerning non-proliferation. It was even seen as example setting for early tangible steps in the pursuit of nuclear disarmament. Such optimism, however, was possibly excessive. It fell victim to the difficult domestic politics in the US, to Israel's imperative need for security and its unique conception of the Middle East, and to Iran's bitter experience of the past four decades. The right wing in US polity could never forget nor forgive the Iranian regime's seizure of the US Embassy in Tehran and holding of American hostages from 1979-80. Israel has always perceived an existential threat in the strong pitch of Ayatollah Khomeini for the Palestinians and his insistent talk, repeated by his successors, that denied the holocaust, and supported wiping out Israel from the world map. Iran's role in the massacre of US Marines in Lebanon in 1984 has always rankled in Washington. Iran, too, on its part has suspected US designs in the humiliating war thrust on it by Saddam Hussein for almost a decade, starting in 1980. There is also anger over infringement of Iran's maritime rights in the Hormuz Strait where a US missile hit Iran's commercial airliner and killed all 270 passengers in 1988. The roots of mutual distrust and grievances, therefore, go deep. No enlightened leadership in any of these countries has so far succeeded in modifying the hardline mindsets in their respective polities.

A precarious situation threatens to return and will persist till acceptable terms are arrived at by the negotiators in Vienna. At present, the US stance conditions its own return to the JCPOA and the lifting of reimposed sanctions on Iran's full compliance with the JCPOA . Iran demands, on the other hand, that the US first remove the sanctions which the Trump Administration imposed from 2018 onwards. Furthermore, there is a shadow of ambiguity on whether and to what extent the US might try to bring in controls on Iran's missiles and regional military role into its decision to revive the JCPOA. Ample promises were made by Joe Biden's team during his presidential campaign, as also after his election, that the nuclear deal with Iran would be back into force. But these are yet to be fulfilled.

Meanwhile, since early 2018, Iran has gradually removed some restrictions placed on its enrichment activities and has denied the IAEA's access to, and monitoring of, locations which are sources of suspicion. Iran's national assembly, the *Majlis*, had enacted a law in December 2020 that ordered the atomic establishment to enrich uranium to

20 per cent and even 60 per cent, specifying quantities too. This law would inhibit the IAEA's monitoring activities and, thus, constrain its ability to certify compliance as prescribed by the JCPOA. The IAEA Director General (DG) visited Tehran in February 2021 to hold talks with the authorities with a view to removing the restrictions placed on IAEA activities by Iran. The new government in Tehran appears to have reasons to distrust the American promises. Some developments in the meanwhile have served to erode mutual trust. Prominent among them has been the killing of a number of Iranian nuclear scientists , the latest being Mohsen Fakhrizadeh in November 2020 by a suspected Israeli covert operation, controlled and guided remotely by, but, as Iran suspected, not without, US complicity. At the time of the newly elected Israeli Prime Minister Bennett's visit to Washington in August 2021, a US official was quoted by the *Wall Street Journal* as asserting: "Since the last Administration left the Iran nuclear deal, Iran's nuclear program has just dramatically broken out of the box, and it's accelerating from week to week."[12]

Iran has more advanced centrifuges and uranium stockpiles as well as technology so that the nuclear "breakout"—the capability to produce a bomb—"is now down to just a few months".

Separately, the IAEA monitors, and recent reports of the IAEA have indicated, Iran's decision to begin producing uranium to levels that surpass the limits stipulated in the JCPOA. It is suspected that Iran might be running the more advanced centrifuges which the JCPOA had shuttered, raising enrichment levels to as high as 60 per cent, and stockpiling this enriched uranium. While as per the IAEA reports, the material is presently in gram quantities, Iran claims it is 25kg. Iran has held in abeyance some aspects of its adherence to IAEA monitoring of its nuclear activities as envisaged by the 2015 deal.

A difficult dialogue between Iranian authorities and the IAEA DG has since tried to resolve these issues. An IAEA Press release of September 12, 2021, averred that after weeks of consultations, Iran had

12. August 26, 2021, 2:52 AM IST. Last updated a day ago.

reached a deal with the IAEA which permitted the inspectors "to service the identified (monitoring and surveillance) equipment and replace their storage media which will be kept under the joint IAEA and AEOI (Atomic Energy Organisation of Iran) seals in the Islamic Republic of Iran."[13]

There is a background to this IAEA Press release. In June 2021, a drone attack reportedly damaged the IAEA's monitoring and surveillance equipment. The Iran Centrifuge Technology Company, located in the city of Karaj, was, as reported in the *New York Times*,[14] on a list of targets that Israel had presented to the Trump Administration in early 2020. The IAEA is now permitted to service cameras in other locations but not at the Karaj site. The DG of the IAEA has been expecting to meet Iran's senior leadership again to discuss disagreements around IAEA nuclear inspections, which he described as being in a "fragile state".[15]

Robert Malley, the US special envoy for Iran, was quoted as asserting that the United States would consider "all options" if Tehran did not consider returning to the constraints of the 2015 deal. He visited the United Arab Emirates, Qatar, and Saudi Arabia to discuss with these US allies the prospects of Iran's return to international negotiations in Vienna. Malley's comments were in sync with US Secretary of State Antony Blinken who also observed in Washington that the US Administration was "prepared to turn to other options if Iran doesn't change course."[16]

---

13. https://www.iaea.org/newscenter/pressreleases/joint-statement-by-the-vice-president-and-the-head-of-atomic-energy-organization-of-the-islamic-republic-of-iran-and-the-director-general-of-the-international-atomic-energy-agency
14. https://www.nytimes.com/2021/06/23/world/middleeast/iran-atomic-agency-attack.html
15. https://www.bloomberg.com/news/articles/2021-10-19/iaea-head-plans-iran-trip-as-allies-seek-to-reboot-nuclear-talks?sref=QmOxnLFz&mkt_tok=MDk1LVBQVi04MTMAAAGAQ4bPRhQ-lsIaB5KdiQAWpUt7DfHy7Q63gVzBEYwLHwdrvIRlDsAR6fUAa3ff2PKaZv7rvbK2IB0tFWCBILx3zBaNmqVBMoZG_JSVS8EwCZc. It was also speculated that the meeting of the IAEA Board of Governors in end November 2021 might assess the situation, and resolve to press Iran to return to compliance.
16. https://mail.google.com/mail/u/0/#inbox/FMfcgzGlkPXMpqxvrRxFTkNVVpzHQksz

The erosion of mutual trust is a huge hurdle on the way to reviving the nuclear deal. The most extensive verification obligations accepted by Iran in the deal would flounder if there is no trust in the other party's commitment to preventing attacks that may be made easy by increased transparency. Particularly destabilising could be attacks against individuals or facilities that are being covered by international inspections and monitoring. Hardline supporters of Israel have always suspected Iran of hiding facilities and maintained the so-called break-out probability. For example, US intelligence had prior knowledge about a secret enrichment plant being built by Iran in Fordow much before Iran shared relevant information with the IAEA in 2009.[17]

The senior officials in the Obama Administration had diligently explained after the deal was finalised, why they had reasons to believe that the implementation of the JCPOA, chapter by chapter, would effectively block the pathways through which Iran could come into possession of a nuclear bomb.

Reimposition of sanctions on Iran made for a double whammy. It cut the ground from beneath the negotiations. It eroded the credibility of the moderate voices in Iran. When former President Rouhani engaged in the talks after his election, his aim was to bring benefits to the masses in Iran. He had the courage to challenge the hardliners during the 2012 elections and even deride the 18,000 plus centrifuges erected by Ahmadinejad as imposing too heavy a price on the Iranian people. Voices of reason at that time had even scaled down the boast about a bomb by asserting that building a few bombs at the cost of earning the enmity and possible nuclear arming of neighbours could be of very limited value.

Attempts at back channel talks with the US ultimately came to naught when then President Trump threw the deal overboard. Rebuilding of trust can be challenging in the absence of a quick return to the *status quo ante*. In the US polity too, next year's Congressional elections might constrain the hands of the negotiators. The question would be whether

17. Burns, n. 8.

the negotiators would come under pressure to show a deal that would be harsher than the JCPOA.

Meanwhile, the burden of regional politics compounded by tense relations among the US, Russia and China makes the political dynamics worse than it was in 2015. Each has an axe to grind in the Middle East conundrum in general, and with Iran in particular.

While all the seven erstwhile parties to the JCPOA appear to be on the same page on resuming talks, the question that hangs in the air is related to the extent to which a revived nuclear accord with Iran will remain free from larger considerations of regional politics as well as the familiar old great power game. Oil also plays a role. Iran's oil output peaked at 2.4 million barrels/day in 2016 following the lifting of sanctions. Thereafter, with reimposition of sanctions, it dived low. Even India had to virtually stop importing Iranian oil. While the resulting spurt in oil price impacts global economic recovery, Organisation of Petroleum Exporting Countries (OPEC) members and the others see oil futures in variance from each other.

When Iran was allied to the West in the time of the Shah's regime, its oil reserves were an asset to stabilise prices, and Iran had few problems with the Gulf countries or Israel. The Shah's ambition to get nuclear power reactors posed no problems. At the time, Iran even had a share in URENCO, the European consortium that undertakes uranium enrichment. All that changed with the Islamic revolution, which demonstrates that geopolitics impacts nuclear issues. As indicated above, the suspicions about Iran's nuclear programme in the 1990s came in handy to put Iran on the spot and the US imposed sanctions on commerce with it.[18]

However, the Islamic revolution in Iran did not alter the Iranians' civilisational self-image of being a sophisticated people aspiring for higher education and technology, alongside their ability to skilfully use statecraft. Numerous Iranians settled in the West are reputed doctors, architects and high class professionals in diverse fields. Recent obituaries of Bani Sadar, Ayatollah Khomeini's longtime associate and Iran's first

18. Obama, n. 10, p. 456.

president after the revolution, recalled his long stay in Paris and his hobnobbing then with renowned intellectuals like Jean Paul Sartre even as he was in close touch with Khomeini. This also shows that for the late Khomeini, things were probably not as one-dimensional as was projected about him. As was evidenced during the revelations of the famous Iran-Contra scandal during the years of the Reagan presidency, Khomeini was not averse to play ball with the US. As William Burns alludes in his memoirs, a grand bargain had been mooted by Iran in 2003 through the Swiss ambassador in Washington—a move that was considered but declined by the then US Administration.[19]

Former Iranian President Hassan Rouhani has mentioned in passing in his memoirs which appeared in 2012 that engaging with the US was like driving a Mercedes Benz car, whereas engagement with Russia or China was akin to driving a Skoda. Iran's self-image was different from that of North Korea. It was keen under Rouhani and Zarif to integrate with the global mainstream. The sunset clauses in the 2015 deal perhaps catered to this probability that an Iran more tied to the world through trade, economy and finance might be that much more averse to flouting the deal. All these predispositions of Iran's elite and expectations of diplomats in the West were severely belied by Trump's rude abrogation of the nuclear deal.

No one is more relieved with the present denouement than Israel which, for its own security, sees Iran as a formidable foe. Israel would want to severely curb Iran's capacity to foment trouble in the Middle East. Israel, therefore, would want Iran to be isolated and squeezed of its natural resources through sanctions. Former Prime Minister of Israel, Benjamin Netanyahu, therefore, went over the shoulders of President Obama to address the US Senate in 2016 to make his case at the invitation of the Republican Party. Saudi Arabia too has been inimical to the regime in Tehran but Crown Prince Mohammad Bin Salman has lately shown that he is not averse to working out a *modus vivendi*. Tehran too has been sending positive messages to Riyadh and other Gulf capitals.

19. Ibid., p. 340.

Iran's hold on Iraq, its clout with Syrian President Bashar Assad, with Hezbollah in Lebanon, the Houthis in Yemen and Hamas in Palestine are seen as destabilising by the US and its allies. But these are regarded as valued assets by the powerful Iranian Revolutionary Guard Corps in Tehran. The highly regarded Commander of the Revolutionary Guards, General Suleimani, was assassinated by the Americans in January 2020. Tehran vowed revenge in public but has left space for returning to the JCPOA.

Iran's guarding its clout may be a reflection of the stark times characterised by desperate diplomatic moves. For example, US President Trump's 2020 pact with the Taliban symbolised unconditional surrender to the Taliban to get the US troops out. Now, post Taliban gaining control of Afghanistan, the US appears inclined to still prefer hope against experience in trying to get the support of the deep state of Pakistan for over-the-horizon operations in Afghanistan. In yet another case, Turkish President Erdogan's shenanigans with the US and NATO show that threats and power play even among alliance partners are becoming more of a norm than an exception. Add to this the tensions in the relations of the US with Russia and China. To make matters worse for the US, even the European allies voice their chagrin at being kept in the dark, the particular case in point being the AUKUS (Australia-UK-US) agreement on the transfer of nuclear-powered submarines.

All said and done, a slide back in diplomatic efforts in Vienna and continuing hiatus with Iran will be in a different setting from 2015 due to so many factors. The ability to mount combined diplomatic pressure on Iran in these circumstances seems less likely, to say the least. Hence, there is much more at stake today in striving for success rather than pushing the envelope for maximal demands. On the specifics of the IAEA role, a breakthrough in the IAEA DG's talks with Tehran may be achievable and will certainly help.

The path, even after revival of the deal, will not be easy. Iran's new regime still remains enigmatic. President Raeesi's address at the UN General Assembly (UNGA) was taciturn about the possibilities of returning to the nuclear deal. It will be necessary for the P5+Germany/

EU to shun unilateral actions and guard the framework constantly against attempts to dismantle it. Removing the sanctions would, as Iran demands, make amends for Trump's unilateral action. The US will need to restore balance by attaching due importance to the commission for consultations prescribed in the JCPOA.

At the moment, Iran's nuclear imbroglio offers little hope of ending. However, continued diplomatic efforts will find a breakthrough only when both sides see common benefit in arriving at a mutually acceptable agreement.

# 8. Nuclear Proliferation in East Asia: Will AUKUS Cause a New Arms Race?

*Jagannath Panda*

The security architecture of Northeast Asia—dominated by questions over the demilitarisation of the Korean Peninsula—has decidedly emerged as a critical geopolitical conundrum requiring the focus of the regional powers in a bid to ensure stability and peace. For democratic 'like-minded' states, China's role as a resident power in Northeast Asia, its ties with North Korea and maintenance of authoritarian strategic influence over the peninsular region amidst Pyongyang's consistent nuclearisation have been the leading factors of concern. In this context, the signing of the Australia, United Kingdom, United States (AUKUS) security partnership in September 2021 has marked a crucial geopolitical development poised to shape the power politics of the Indo-Pacific region and Asian subcontinent.[1] The "enhanced trilateral security partnership" will also play a major role in the Northeast Asian security architecture—with a focus on nuclear proliferation, and the potential onset of a never-ending arms race in an ironic battle to curb the same.

Pyongyang was quick in voicing its opposition to the AUKUS pact with an official from the Foreign Ministry stating that the deal has the capability to "upset the strategic balance in the Asia-Pacific region".[2] The official further threatened counter-measures should the "undesirable and

1. "Joint Leaders Statement on AUKUS", The White House, September 15, 2021. https://www.whitehouse.gov/briefing-room/statements-releases/2021/09/15/joint-leaders-statement-on-aukus/
2. "Aukus Could Trigger a 'Nuclear Arms Race', says North Korea", *BBC*, September 20, 2021. https://www.bbc.com/news/world-asia-58621056

dangerous" deal lead to national security ramifications for North Korea.[3] The reaction by North Korea echoed that of its "Treaty of Friendship, Cooperation and Mutual Assistance"[4] partner, China, whose Foreign Ministry spokesperson Zhao Lijian termed AUKUS a pact that would "gravely undermine regional peace and stability, aggravate the arms race and impair international nuclear non-proliferation efforts".[5] With both North Korea and China closely watching AUKUS related developments, it becomes all the more clear that America's changing—and solidifying—strategic reorientation towards Asia and the Indo-Pacific is a growing cause of worry for the two powers. Ultimately, there are key factors that showcase how AUKUS will fit into, and shape, the Northeast Asian security outlook, the South and North Korea divide, and China's role in the region. This chapter evaluates the recent trends and perceptions of the major powers on proliferation and arms race debates in the region.

## Unfolding of a New (Nuclear) Arms Race

AUKUS' creation—which came as a rather unexpected announcement—is driven by "enduring ideals and shared commitment to the international rules-based order"[6] of the three democratic Anglo powers. The trilateral security partnership comes as a defence-oriented grouping, with a broad and wide focus on the "investment, research and innovation, collaboration, and adoption" of advanced and critical military technologies, to give them a crucial edge as they navigate an

3. Chaewon Chung, "North Korea Slams AUKUS Deal as 'Undesirable and Dangerous'", *NK News,* September 20, 2021. https://www.nknews.org/2021/09/north-korea-slams-aukus-deal-as-undesirable-and-dangerous/
4. Khang Vu, "Why China and North Korea Decided to Renew a 60-Year-Old Treaty", *The Interpreter*, July 30, 2021. https://www.lowyinstitute.org/the-interpreter/why-china-and-north-korea-decided-renew-60-year-old-treaty
5. "Foreign Ministry Spokesperson Zhao Lijian's Regular Press Conference on September 22, 2021", Embassy of the People's Republic of China in the United States of America, September 22, 2021. https://www.mfa.gov.cn/ce/ceus/eng/fyrth/t1908814.htm
6. See n. 1.

uncertain security environment in the region.[7] As modern warfare comes to be characterised by hybrid threats magnified by technological disruptions, AUKUS is set to be a pivotal force to possibly counter them.[8] However, beyond such implications, the most prominent feature of the AUKUS trilateral was its first initiative, which sought to build on the three states' "common tradition as maritime democracies" to leverage the capabilities of the US and the UK, and help Australia build nuclear-powered submarines for the Royal Australian Navy.[9] While the initial plan is to find an optimal pathway to deliver this within 18 months, the overall ambition to bring nuclear-powered submarine capacity to Australia—which has negligible experience with nuclear technology—is likely to be a decades-long project and will invariably and deeply intertwine both Washington and London with the Indo-Pacific region. In this context, AUKUS is posed to be a permanent fixture with a concentrated focus on military-technological partnerships while its operational impact will be trans-sectoral. Hence, AUKUS' objectives are contextual and cross-continental, and will impact Asia as a whole. In this regard, the partnership has been viewed very obviously as a reaction to the debased security climate in the region, the factors for which are varied. The arrival of AUKUS is the latest key variable denoting the onset of a 'nuclear arms race' in Northeast and East Asia.

Before moving to the factors shaping the tense regional security climate, it also becomes imperative to study them in the context of the critical conventions of current peacekeeping systems which form the framework for implementing the rules and norms regarding non-

7. Brendan Nicholson, "Morrison says AUKUS will Strengthen Cooperation on Critical Technologies," *The Strategist*, Austrian Strategic Policy Institute, November 17, 2021. https://www.aspistrategist.org.au/morrison-says-aukus-will-strengthen-cooperation-on-critical-technologies/.
8. Fergus Hanson and Danielle Cave, "The Real Potential of AUKUS is About far More Than Submarines," *The Strategist*, Austrian Strategic Policy Institute, September 20, 2021. https://www.aspistrategist.org.au/the-real-potential-of-aukus-is-about-far-more-than-submarines/.
9. n. 1.

proliferation.[10] The key international agreements governing state behaviour vis-à-vis the development and use of nuclear weapons are the Treaty on Non-Proliferation of Nuclear Weapons (NPT), the Comprehensive Test Ban Treaty (CTBT), the Fissile Material Cut-off Treaty (FMCT), the International Atomic Energy Agency (IAEA) safeguards framework, and different export control arrangements. Each of these treaties requires closer critical study to manage its deficits. For instance, with the NPT[11], attention must be called to the fact that the focal principle of the arrangement—that Nuclear Weapon States (NWS) would denuclearise if the Non-Nuclear Weapon States (NNWS) agreed to not secure atomic weapons—has been adversely affected by the general absence of progress regarding disarmament. In this context, both North Korea and China –in similar (if not the same) terminology –have claimed that AUKUS could cause a 'nuclear arms race' in Asia. They remain anxious that the arrangement could lead to an increased number of nuclear-powered general-purpose attack submarines (SSNs) among other weaponry. Amidst such motives, a large part of the focus on AUKUS has covered its goal of getting atomic power submarines (as of yet without atomic weapons) for the Royal Australian Navy. However, with regards to AUKUS, the argument presented in favour of non-violation of the NPT is that the treaty does not stop the NNWS from attaining non-explosive military uses of nuclear products, such as naval propulsion reactors. With the AUKUS submarines being nuclear-powered and not nuclear-weaponised, the imposition of the treaty remains beyond the purview of AUKUS, protecting the tri-powers' individual commitment to de-nuclearisation.[12] In other words, as the AUKUS joint leaders' statement

10. "Nuclear Weapons Challenges In Asia", Asia-Pacific Centre for Security Studies, April 22, 2000. https://apcss.org/Publications/Nuclear%20Weapons%20Challenges%20in%20Asia.html
11. Manseok Lee and Michael Nach, "Challenges to the Nuclear Non-Proliferation Treaty", *Strategic Studies Quarterly*, AIR University, 2020. https://www.airuniversity.af.edu/Portals/10/SSQ/documents/Volume-14_Issue-3/Lee.pdf
12. John Carlson, "AUKUS Nuclear-Powered Submarine Deal—Non-Proliferation Aspects", *APLN*, September 17, 2021. https://www.apln.network/analysis/commentaries/aukus-nuclear-powered-submarine-deal-non-proliferation-aspects

argues, the trilateral nuclear-powered submarine initiative will not hinder Australia's commitments and obligations as an NNWS, including those presented via membership in the International Atomic Energy Agency (IAEA), and the three states' leadership role in bringing about global non-proliferation.

Such leeway in the AUKUS arrangement, while ensuring its authenticity, leads to concerns regarding the nuclear arms race in the region. For North Korea and China, AUKUS becomes more worrisome as the US by precedence has not imparted its valued atomic submarine technology before—citing domestic laws—to even its nearest partners, such as India, other Five Eyes intelligence sharing partners like Canada and New Zealand, or treaty security partners like South Korea.[13] Australia has emerged as an exception to this; even as this has been at the expense of trans-Atlantic ties by angering France due to the withdrawal of an Australian arrangement with Paris for building conventional submarines.[14] This move has greatly upset trans-Atlantic ties; to ensure the longevity of its decision at such a high cost, the US' focus on AUKUS is most unlikely to waver in the near future.

Keeping this narrative in mind, Korean experts have begun arguing that the US could now contemplate sharing its nuclear-powered submarine technology with Seoul as well[15], especially considering that it is a coveted capability which South Korea has sought for long as a means of keeping North Korea's ballistic and nuclear adventurism in check. For South Korea, gaining such nuclear expertise has been an enduring dream—but something that the Republic of Korea's (ROK's) treaty ally, the US, has steadfastly opposed. Therefore, despite its ambition, any

13. Sandeep Unnithan, "Why the US Won't Give India Nuclear Submarines", *India Today*, September 20, 2021. https://www.indiatoday.in/india-today-insight/story/why-the-us-won-t-give-india-nuclear-submarines-1854818-2021-09-20
14. "AUKUS Fiasco: France Accuses Australia, US of 'Lying' in Escalating Crisis", *The Economic Times*, September 22, 2021. https://economictimes.indiatimes.com/news/international/world-news/aukus-fiasco-france-accuses-australia-us-of-lying-in-escalating-crisis/articleshow/86334482.cms?from=mdr
15. Peter K. Lee, "AUKUS and South Korea's Dilemma", *The Korea Times*, October 4, 2021. https://www.koreatimes.co.kr/www/opinion/2021/11/197_316305.html

partnership like AUKUS with the US to acquire nuclear technology capabilities has not been an option. In fact, according to terms set by the 1972 civil nuclear cooperation agreement between the US and South Korea, although both countries share robust cooperation in the civil nuclear sector,[16] Washington has been reluctant to provide consent for the use of nuclear materials for military purposes, even when the agreement was renegotiated and updated in June 2015.[17]

In other words, while acquiring nuclear submarines has been a matter of interest and discussion in Seoul for 20 years or so, AUKUS' establishment has reinvigorated it.[18] The fact that it faces a growing threat from North Korea in the maritime domain—with its northern neighbour testing several submarine-launched ballistic missiles over the past few years[19]—has become a catalyst for South Korea's ambitions. According to the Nuclear Threat Initiative (NTI), a US-based non-profit organisation, Seoul has an advanced nuclear programme and is capable of building its own nuclear-propelled naval vessels (both submarines and surface);[20] however, for such equipment to be operational, Seoul will need access to enriched uranium fuel—for which Washington has not yet provided

16. See Robert Einhorn, "US-ROK Civil Nuclear Cooperation Agreement: Overcoming the Impasse," speech, Brookings, October 11, 2013. https://www.brookings.edu/on-the-record/u-s-rok-civil-nuclear-cooperation-agreement-overcoming-the-impasse/.
17. See Bureau of International Security and Nonproliferation, "US-Republic of Korea (ROK) Agreement for Peaceful Nuclear Cooperation," US Department of State. Accessed on December 19, 2021. https://www.state.gov/remarks-and-releases-bureau-of-international-security-and-nonproliferation/u-s-republic-of-korea-r-o-k-agreement-for-peaceful-nuclear-cooperation/.
18. Choe Sang-Hun, "South Korea Has Long Wanted Nuclear Subs: A New Reactor Could Open a Door," *The New York Times,* December 13, 2021. https://www.nytimes.com/2021/12/13/world/asia/south-korea-nuclear-submarines.html.
19. Choe Sang-Hun, "North Korea Tests First Submarine-Launched Missile in Two Years," *The New York Times*, October 18, 2021. https://www.nytimes.com/2021/10/18/world/asia/north-korea-ballistic-missile-test-washington.html.
20. George M. Moore, Cervando A. Banuelos, and Thomas T. Gray, *Replacing Highly Enriched Uranium in Naval Reactors* (Washington DC: Nuclear Threat Initiative, March 2016),. https://media.nti.org/pdfs/Replacing_HEU_in_Naval_Reactors_Report_FINAL.pdf?utm_source=pocket_mylist.

"so-called advance consent".[21] The formation of AUKUS comes as an indication that the US is more concerned than ever with the security dynamics of the region vis-à-vis China, and may now be willing to revisit its long-held policies on the matter to consider becoming a partner in South Korea's nuclear submarine initiative.

For North Korea, which perceives its nuclear capacity as critical for protecting its national security interests, AUKUS' formation has accentuated Pyongyang's fears since Seoul now has the option of turning not just to Washington, but also Paris, for attaining nuclear-powered technology. France and South Korea have, over the past three years, sought to increase defence cooperation, focussing on building technological cooperation and economic ties.[22] With the French military industry reeling from the cancellation of the US$ 66 billion[23] submarine deal with Australia, it could potentially welcome Korean collaboration. Notably, France has several decades of experience of building nuclear submarines; the French fleet includes four ballistic missile submarines (SSBNs)—the Le Triomphant class submarines, which are set to be replaced by upgraded third generation sub-surface vessels beginning 2035—and seven nuclear-powered attack submarines (SSNs)—the Barracuda (or Suffren) class submarines, which are set to gradually replace the existing Rubis and Amethyste class vessels.[24] While the US'

21. Daniel Horner, "S. Korea, US Sign Civil Nuclear Pact," Arms Control Association, July/August 2015. https://www.armscontrol.org/act/2015-07/news/s-korea-us-sign-civil-nuclear-pact.
22. "S. Korea, France Agree to Bolster Defence Cooperation", *Yonhap News Agency*, April 18, 2019. https://en.yna.co.kr/view/AEN20190418009600325
23. Kyle Mizokami, "France Is Furious with Australia Over a Broken Submarine Deal—Here's Why", *Popular Mechanics*, September 28, 2021. https://www.popularmechanics.com/military/navy-ships/a37667312/france-australia-broken-submarine-deal-explained/
24. "France Submarine Capabilities," Submarine Proliferation Resource Collection, Nuclear Threat Initiative, September 27, 2015, updated September 21, 2021. https://www.nti.org/analysis/articles/france-submarine-capabilities/. Also see Ryan Pickrell, "France is Furious after Australia Scrapped a Submarine Deal to Partner with the US and UK. Here's how American, British, and French Subs Stack up," *Business Insider*, September 22, 2021, https://www.businessinsider.in/international/news/france-is-furious-after-australia-scrapped-a-submarine-deal-to-partner-with-the-us-and-uk-heres-how-american-british-and-french-subs-stack-up/articleshow/86412810.cms.

Virginia class submarines can boast of superior capabilities, the French Barracuda class submarines are armed with the most advanced, world-class torpedoes and anti-ship missiles, and use low-enriched uranium. Seoul recently announced plans to establish an indigenous nuclear reactor that could potentially take forward South Korea's nuclear submarine programme;[25] France is already working with Brazil on a similar project to help with the design and integration of nuclear submarines.[26] The Suffern class submarines have a diameter only slightly larger than the ROK's KSS-III class, and should the KSS-III be the basis for its nuclear submarine (KSS-N), France would be an ideal primary partner for Seoul. While both Paris and Seoul have previously mentioned the possibility of collaboration in this sector, AUKUS (and the changing regional security dynamics vis-à-vis China) could reinforce this idea and actualise practical cooperation between them. At the same time, South Korea's and Australia's 2+2 meetings would be closely watched by North Korea.[27] Even though these meetings have not resulted in any concrete action steps being committed to by both countries, there seems to be a modest pace at which the relations between the two are moving, especially vis-à-vis defence procurement.

## Beyond AUKUS: Recent Trends in Northeast and East Asia

The Korean Peninsula– and the broader Northeast Asian geography—as well as the Indo-Pacific are two of the world's more contentious and sensitive regions. With new developments in the Indo-Pacific, such as

25. Eunhee Lee, "원자력연구원, "감포 문무대왕과학연구소에 해양용 ARA 연구로 건설 [Translated: Atomic Energy Research Institute, "Construction of ARA Research Reactor for Marine Use at Gampo Munu Daewang Science Research Center]," *Newsis 20*, November 10, 2021. https://newsis.com/view/?id=NISX20211110_0001646864.
26. H I Sutton, "South Korea's First Nuclear Submarine Looks Closer," *Naval News*, December 15, 2021. https://www.navalnews.com/naval-news/2021/12/south-koreas-first-nuclear-submarine-looks-closer/.
27. Bill Paterson, "AUKUS's Implications for Australia–South Korea Defence Collaboration", *The Strategist*, September 29, 2021. https://www.aspistrategist.org.au/aukuss-implications-for-australia-south-korea-defence-collaboration/

the Quadrilateral Security Dialogue (Quad) and AUKUS, how these mechanisms expand their focus beyond the scope of Asian power politics remains to be seen. North Korea's continued nuclear and missile tests, China's proliferation promotion via domestic manufacturing, a growing focus on the underwater arms race, with rapid advances in the navies of the Asian states, Japan's consistent defence budget increases, and South Korea's quest for nuclearisation are factors that have greatly impacted the proliferation trends in the region. Northeast Asia is home to multiple states with atomic weapons, two alliances that join broadened atomic deterrence, increasingly muddled and tense relations between the United States and China, and the ever-growing North Korean nuclear abilities.[28] Such circumstances place an impressive strain on the ROK-US and Japan-US treaty alliances to develop a dialogue and collaborate on regional initiatives to deal with these dangers.

Chief amongst these threats is China's rapidly accelerating nuclear programme, which has only grown and expanded with the intensifying Sino-US rivalry. For instance, China at present, has 350 atomic warheads, a stockpile that has consistently expanded. It has, at the same time, tried to modernise and develop its nuclear delivery frameworks, which incorporate more than 143 nuclear land-based missiles and a predetermined number of submarines, Submarine-Launched Baltistic Missiles (SLBMs), and bombers.[29] China's self-expressed atomic strategy has been to keep its abilities at the base level needed to keep up with its national security and to dissuade a possible first strike, being a "no first use" strategy propagator. Yet, it should be noted that according to the Pentagon, an assessment of China's production capacity suggests that it could have as many as 1,000 nuclear warheads within a decade (by 2030) as it moves towards its goal of modernising its warfare capabilities by

28. Simon Duke, "Northeast Asia and Regional Security", *The Journal of East Asian Affairs,* 9 (2), pp. 323–384, 1995. http://Www.Jstor.Org/Stable/23253997.
29. "Arms Control and Proliferation Profile: China", Arms Association. https://www.armscontrol.org/factsheets/chinaprofile

2027 and transforms the People's Liberation Army (PLA) into a 'world class military' by 2049.[30] This would be inclusive of 700 "deliverable" warheads by 2027, which can be mounted onto existing missiles without any delay.[31] While this will still be significantly dwarfed by the US' own capabilities (comprising a massive stockpile of 3,750 atomic warheads) and Russia's—both of which own the majority of the world's total nuclear weapons—it nevertheless marks a rather sharp escalation in Beijing's nuclear capacity. Such build-up comes as part of China's larger campaign for military modernisation to bolster its defence planning vis-à-vis the US. This includes inducting advanced military technologies (a core area of focus under AUKUS) that drive an arms race for Artificial Intelligence (AI),[32] and weapon systems such as hypersonic missiles.[33] For instance, in August 2021, China tested a nuclear-capable hypersonic missile that circled the globe before hitting its target. The test demonstrated China's

30. Office of the Secretary of Defence, *Military and Security Developments Involving the People's Republic of China*, Annual Report to Congress (Washington DC: Department of Defence, 2021). https://media.defence.gov/2021/Nov/03/2002885874/-1/-1/0/2021-CMPR-FINAL.PDF; Amanda Macias, "China is Rapidly Expanding its Nuclear Arsenal, Pentagon says in New Report," *CNBC*, November 3, 2021. https://www.cnbc.com/2021/11/03/china-is-rapidly-expanding-its-nuclear-arsenal-pentagon-says.html.
31. Peter Martin and Anthony Capaccio, "China's Nuclear Arsenal is Growing Faster Than Expected, Pentagon Says," Bloomberg, November 3, 2021. https://www.bloomberg.com/news/articles/2021-11-03/pentagon-sees-china-nuclear-arsenal-growing-faster-than-expected.
32. Julian E. Barnes and Josh Chin, "The New Arms Race in AI," *Wall Street Journal*, March 2, 2018. https://www.wsj.com/articles/the-new-arms-race-in-ai-1520009261.
33. Hypersonic missiles are simply defined as "manoeuvrable weapons that can fly at speeds in excess of Mach 5, five time the speed of sound". See Dinakar Peri, "Explained: The Arms Race Towards Hypersonic Weapons," *The Hindu*, December 7, 2021. https://www.thehindu.com/sci-tech/technology/explained-the-arms-race-towards-hypersonic-weapons/article37874968.ece. Such weapons boast of several advantages compared to traditional ballistic missiles; they fly at comparatively lower altitudes (through low orbit space) and can be more easily manœuvred, making them difficult to detect and, therefore, intercept. Also see "China's Hypersonic Missile Test: All You Need to Know," *The Times of India*, November 22, 2021. https://timesofindia.indiatimes.com/world/china/chinas-hypersonic-missile-test-all-you-need-to-know/articleshow/87853547.cms.

incredible progress in hypersonic technology as well as its advanced space capabilities.[34] Notably, the test caught the US intelligence agencies by surprise in a "Sputnik moment" for America—pointing to the fact that as China perfects this technology, it will be able to launch nuclear warheads over the US that evade detection by American anti-missile systems situated in the northern hemisphere.[35] Even though China argues that its threat as a military, or even nuclear, power is imagined—particularly considering Beijing's public commitment to "no first use"—such expanding capabilities have put Washington on edge and all but guaranteed a new arms race in advanced weaponry and military equipment.

Furthermore, China's growing revisionism—especially vis-à-vis the Korean Peninsula—along the lines of its national interest driven outreach puts a strain on its commitment to denuclearisation. For instance, while it was the Democratic People's Republic of Korea's (DPRK's) nuclear development that marked the first major signs of strain on China-North Korea ties[36], the two have successfully renewed their Friendship Treaty despite their differences on nuclearisation. Such movements have highlighted to the democratic 'like-minded' states that China's main goal in the Korean Peninsula remains one-dimensional. China's advances towards North Korea have been largely limited, with Beijing looking to influence Pyongyang bilaterally rather than through multilateral mediums. For China, the treaty remains indispensable as it has played a key role in maintaining China's engagement in the Korean Peninsula's peace over the years. Beijing believes that South Korea and the US wish to dominate the peninsula, excluding China from the narrative.

In this context, regional players such as Japan and South Korea have begun charting even more active non-proliferation trends of their own. Japan's 2021 White Paper has highlighted North Korea as

---

34. Demetri Sevstopulo and Kathrin Hille, "China Tests New Space Capability with Hypersonic Missile," *Financial Times,* October 17, 2021. https://www.ft.com/content/ba0a3cde-719b-4040-93cb-a486e1f843fb.
35. Martin and Capaccio, n. 31.
36. "The China–North Korea Relationship", Council on Foreign Relations, June 25, 2019. https://www.cfr.org/backgrounder/china-north-korea-relationship

an "imminent threat"[37]; Tokyo's growing defence budget has kept such security threats in mind. Post North Korea's recent nuclear tests, and the growing Chinese incursions in the Taiwan Strait as well as East China Sea (ECS), Japan even approved a record high extra defence budget of US$ 6.7 billion in November 2021[38], adding to its already incremented 2020 defence budget. In this context, even as AUKUS does not limit its impact to 'China containment' but rather aims at advancement of a democratic, rules-based order across Asia while proving to be a challenge to both Beijing and Pyongyang, the role of actors like Japan with a focus on building their own national security, will grow in the coming years. AUKUS must also be viewed as a more subscripted version of the Five Eyes (FVEY), the world's key intelligence sharing alliance, based on both being initiatives of the Anglo-sphere democratic nations. FVEY (the US, the UK, Australia, Canada and New Zealand) and AUKUS already share foundational members; the potential to expand in the future to include more states for AUKUS remains a real possibility. In this regard, FVEY has reportedly[39] worked for long with unofficial partners like France, Japan and South Korea on the North Korean threat. Representatives from all eight countries possibly met in January 2020 to discuss the same.

Furthermore, the growing number of underwater nuclear and non-nuclear vessels portends the onset of a new underwater arms race in the region.[40] AUKUS could reasonably support a future wherein Seoul joins

37. "Japanese Defence White Paper Highlights 'North Korea's Imminent Threat'", *NK News*, July 12, 2021. https://www.nknews.org/2021/07/japanese-defence-white-paper-highlights-north-koreas-imminent-threat/
38. Kosuke Takahashi, "Japan Approves Record Extra Defence Budget", *The Diplomat*, November 26, 2021. https://thediplomat.com/2021/11/japan-approves-record-extra-defence-budget/
39. "'Five Eyes' Intel Alliance Ties up with Japan on North Korea threat", *The Japan Times*, January 27, 2020. https://www.japantimes.co.jp/news/2020/01/27/national/five-eyes-intelligence-sharing-alliance-partners-japan-north-korea/
40. Liu Zhen, "The Underwater Arms Race: China, Aukus and a Deepening Submarine Rivalry", *South China Morning Post*, October 11, 2021. https://www.scmp.com/news/china/military/article/3151929/underwater-arms-race-china-aukus-and-deepening-submarine

the underwater nuclear race by turning to democratic allies.[41] Yet, such a nuclear underwater arms race may just be feasible for those Indo-Pacific powers that have huge maritime capacities and, essentially, the vital finances to help such acquisitions. Northeast Asia has witnessed many arms race admonitions over the decades, such as Australia's acquisition of the nuclear weapons-capable F-111 and South Korea's Technical High Attitude Area Defence (THAAD) deployment.[42] Be that as it may, on account of AUKUS, nuclear-fuelled submarines are not a breakthrough advancement of capacity for the Indo-Pacific. As mentioned earlier, Beijing already has an armada of atomic-fuelled submarines and is expanding it. India too is well in the game with its first indigenously developed atomic submarine. The arms race contest is frequently a "kneejerk"[43] response whenever new arms procurement surfaces, regardless of whether it is in the form of expansion of capability to new states or increase in the existing stockpiles of states already possessing them. The issue is that 'race' suggests a hard and fast challenge where no less than one party looks to succeed. There is no such uniform peculiarity across the Indo-Pacific region, which comprises different sub-areas, each with its own kind of economic, socio-cultural and political settings.

Northeast Asia would lead such an arms race in the Indo-Pacific. Beijing would expand its maritime capability, paying little heed to AUKUS. The agreement only gives it more motivation to seek further military development. South Korea only recently declared its entrance into the elite submarine-launched ballistic missile club[44] in the wake of North Korea's advancement of such a capacity, and there are domestic

41. Jagannath Panda, "What AUKUS Means to North Korea", *38North*, December 9, 2021. https://www.38north.org/2021/12/what-aukus-means-to-north-korea/
42. Brendan Taylor and David Envall, "A Nuclear Arms Race in Northeast Asia?", *Asia and the Pacific Policy Society*, December 13, 2017. https://www.policyforum.net/nuclear-arms-race-northeast-asia/
43. Collin Koh, "No AUKUS Arms Race in Southeast Asia", *East Asia Forum*, November 2, 2021. https://www.eastasiaforum.org/2021/11/02/no-aukus-arms-race-in-southeast-asia/
44. "South Korea Conducts Submarine-Launched Ballistic Missile Test", *Naval News*, July 4, 2021. https://www.navalnews.com/naval-news/2021/07/south-korea-conducts-submarine-launched-ballistic-missile-test/

discussions on whether Seoul now needs atomic-fuelled submarines for more credible deterrence. Japan has not abandoned the choice of procuring such assets either, especially focussed on strike capabilities, also in the face of North Korea's SLBM launch (therein building a small example of an arms race concept, with both the ROK's and Japan's outlook stemming from North Korea's actions).[45] Furthermore, India looks set to increase its atomic armada—a choice that was long in the pipeline even before AUKUS. In this regard, the 'weapons contest' conceived by the pundits of AUKUS would not be restricted to atomic submarines or atomic weapons. This dynamic would be all the more expansive, stretching out into different fields of conventional arms as well.

## Drawing Implications for India and the Indo-Pacific

In essence, the nuclear nuances and heightening tensions in the East Asian region point to an emerging, if not intensifying, trend of proliferation. While North Korea is an obvious point of concern in this regard, it is the US-China competition and conflict that is spurring an arms race in the region. Most notably, such a nuclear race amongst the great powers competing for primacy has forced the middle power democratic states—Japan and South Korea—to revisit and recalibrate their own stances in the light of external pressures on their national security interests by rival states in the region.[46] Interestingly, although it is an NNWS, Japan possesses "nuclear latency"[47] as it holds a massive

45. "North Korea's New SLBM Prompts Japan to Consider Seeking Strike Capability", *The Japan Times*, October 20, 2021. https://www.japantimes.co.jp/news/2021/10/20/national/north-korea-japan-strike-capability/
46. Hart Rapaport, "Nuclear Proliferation Concerns in East Asia: Beyond North Korea," *The Diplomat,* November 26, 2021. https://thediplomat.com/2021/11/nuclear-proliferation-concerns-in-east-asia-beyond-north-korea/.
47. Nuclear latency refers to the capacity to develop nuclear weapons and is often associated with providing benefits of deterrence in conflicts. See Mathew Fuhrmann and Benjamin Tkach, "Almost Nuclear: Introducing the Nuclear Latency Dataset," *Conflict Management and Peace Science,* 32, No. 4 2015, pp. 443-461. https://doi.org/10.1177/0738894214559672; Rupal N Mehta and Rachel Elizabeth, "The Benefits and Burdens of Nuclear Latency, *International Studies Quarterly*, 61, No. 3 September 2017, pp. 517-528. https://doi.org/10.1093/isq/sqx028.

stock of plutonium—a full nuclear fuel cycle—as well as advanced industries, and scientific and technical expertise, capable of producing weapons of mass destruction—which has long been a source of concern for advocates of non-proliferation.[48] Despite Japan's aversion to nuclear weapons, owing to its history, as Tokyo revisits its pacifist Constitution to better safeguard its security interests in a region in flux, the worries are that it could resort to nuclear weapons development as peace and stability in the region are disrupted. In South Korea, the discourse on developing nuclear-powered submarines is already well underway with the recent announcement of the establishment of a reactor.

Simply put, nuclear weapons precipitate nuclear weapons and China's and North Korea's massive—and still growing—nuclear capability looms as an imminent threat for Japan, South Korea and the East Asian region at large, both in terms of their traditional security considerations and their psyche.[49] Almost half of the South Korean population supports developing nuclear weapons as a step towards safeguarding the nation from North Korea's nuclear threat amid a Mutually Assured Destruction (MAD) thinking.[50] While the Japanese public is often said to have a pronounced "nuclear allergy"[51]—an opinion poll in 2017 found almost 70 per cent of the Japanese public in favour of remaining non-nuclear[52]—

48. See Stephen Herzog, "The Nuclear Fuel Cycle and the Proliferation 'Danger Zone'," *Journal for Peace and Nuclear Disarmament* 3, No. 1, 2020, pp. 60-86, DOI: 10.1080/25751654.2020.1766164; Alexandre Debs and Nuno P. Monteiro, "Cascading Chaos in Nuclear Northeast Asia," *The Washington Quarterly*, 41, No. 1, 2018, pp. 97-113, DOI: 10.1080/0163660X.2018.1445902; "Japan," Nuclear Threat Initiative, October 21, 2021. https://www.nti.org/countries/japan/.
49. Ibid.
50. Morten Soendergard Larsen, "Talk of a Nuclear Deterrent in South Korea," *Foreign Policy,* September 9, 2021. https://foreignpolicy.com/2021/09/09/south-korea-nuclear-deterrent-north-korea/.
51. Yasumasa Tanaka, "Japanese Attitudes Toward Nuclear Arms," *Public Opinion Quarterly* 34, No. 1, Spring 1970, pp. 26-42. https://doi.org/10.1086/267770.
52. Genron NPO, and East Asia Institute, *The 5th Japan–South Korea Joint Public Opinion Poll (2017): Analysis Report on Comparative Data* (Tokyo: Genron NPO, July 2017). http://www.genron-npo.net/en/archives/170721_en.pdf.

and a profound commitment to the three non-nuclear principles (non-possession, non-production, and non-introduction). Yet, fears that this may change, persist. The fact that Tokyo remains conspicuously absent from the Treaty on the Prohibition of Nuclear Weapons (TPNW)—which was first proposed a decade ago and came into effect in January 2021, and is recognised as the "beginning of the end of the nuclear age"—demonstrates Japan's continued reliance on the US' nuclear umbrella for its security.[53] In this context, Japan's renunciation of the TPNW has come to challenge the country's commitment to a nuclear weapon-free world.[54] In other words, for Japan, its alliance with the US provides it with nuclear deterrence and, therefore, forms a critical and necessary strategic option to navigate regional challenges. Under such conditions, should US-Japan relations undergo tensions, Tokyo could very well be forced to consider developing its own nuclear weapons capabilities.

Considering such trends, the strategic issues and nuclear politics of East Asia are of extreme importance to New Delhi, in terms of both its Act East policy and its Indo-Pacific outlook. India already shares a complex relationship with China owing to their highly tense border dispute and differences on matters like Tibet. While India too is a nuclear state, China has, thus, far refused to recognise it as such—pointing to their complicated nuclear relationship as well. Post the India-China Galwan Valley clash of June 2020, as both nuclear-armed states became involved in a military standoff, India conducted a series of missile tests and Beijing fired ballistic missiles in the South China Sea (near the Spratly and Paracel Islands). China's nuclear posture is driven primarily by its threat perception of the US; Beijing's internal discussions on expanding its nuclear arsenal, rethinking its no first use policy, and developing and deploying advanced technology weapons (like hypersonic

53. Sayuri Romei, "Nuclear Ban Treaty Offers Rare Chance for Japan," *The Asahi Shimbun,* April 28, 2021. http://www.asahi.com/ajw/articles/14330097.
54. For a detailed discussion of Japan and the TPNW, see Jonathon Baron, Rebecca Davis Gibbons and Stephen Herzog, "Japanese Pubic Opinion, Political Persuasion, and the Treaty on the Prohibition of Nuclear Weapons," *Journal for Peace and Nuclear Disarmament,* 3, No. 2, 2020, pp. 299-309, https://doi.org/10.1080/25751654.2020.1834961.

missiles) are aimed at enhancing China's nuclear deterrence with respect to the US.[55] Nevertheless, this growing arms race has an acute impact on India's strategic thinking and security calculus. Beijing presents a serious threat for India and in view of its growing conventional, nuclear and advanced technology capabilities, it has become vital for New Delhi to pursue both internal and external balancing strategies.[56] Greater outreach to, and collaboration with, democratic partners in East Asia, that face similar dynamics, can be a crucial strategy.

## Summing Up

Despite being within the purview of the NPT, and not being responsible entirely for a new arms race, AUKUS has indeed impacted greatly the future of non-proliferation efforts in the Asian security landscape. As Rafael Grossi, the director general of the International Atomic Energy Agency (IAEA), has highlighted, the pact could set a new precedent for nuclear-powered submarines in the region, which would ultimately only further the militarisation race in the region.[57] The IAEA fills in as the world's watchman on the question of nuclearisation, setting up safeguards against development of atomic weaponry; as such, it is the association that keeps the world 'safe' from a possible nuclear catastrophe. The most recent development in the IAEA's roster is AUKUS, which could start a risky trend that could eventually hurt future denuclearisation endeavours.[58] Essentially, for

55. Manpreet Sethi, "The Good, the Bad, and the Ugly in India-China Nuclear Relations," Institute of Peace and Conflict Studies, October 27, 2020. http://www.ipcs.org/comm_select.php?articleNo=5734.
56. Rajesh Basrur, "The India-China Nuclear Dynamic: India's Options," Observer Research Foundation, December 22, 2020. https://www.orfonline.org/research/india-china-nuclear-dynamic-india-options/.
57. "IAEA Chief: Aukus Could set Precedent for Pursuit of Nuclear Submarines", *The Guardian*, October 20, 2021. https://www.theguardian.com/world/2021/oct/19/iaea-aukus-deal-nuclear-submarines
58. Lara Geiger, "The AUKUS Submarine Deal Sinks the Prospect of Long-Term Non-Proliferation", *Columbia Political Review*, November 8, 2021. www.cpreview.org/blog/2021/11/the-aukus-submarine-deal-sinks-the-prospect-of-long-term-non-proliferation

Australia to operate atomic controlled submarines, it will need to become the principal non-nuclear weapon state to use a provision that permits it to remove nuclear material from the safeguards arrangement of the IAEA. AUKUS, hence, marks a change in the structural plates of the international strategy on non-proliferation: it compromises the capacities of the IAEA to check atomic weapons development.[59] Such loopholes can be used in the future as well, by states like China and North Korea.

Moving forward, while the AUKUS trilateral does hold risks, it also promises to undertake deep data and technological sharing, further defence innovation, industrial growth and supply chains, improve digital connectivity, artificial intelligence, quantum advancements and undersea capacities. AUKUS' potential collaboration with the Quad—US, Australia, Japan, India –can also be envisioned via such dimensions. For countries like India, AUKUS does not resolve major questions like foreign policy driven interference, economic coercion, financial pressure and human rights which are relevant to the great power competition between China and the US and its allies. Be that as it may, India recognises that it will improve the Indo-Pacific region's military and technical capacities, regardless of the fact that the conceived eight submarines will require many years to become operational. In addition, AUKUS will be commonly gainful in that the militarily overstretched US needs its partners to assume a greater role in the region; this remains a major factor behind India and powers like Japan welcoming the initiative even though they were left out of the negotiation process.

At the same time, Foreign Secretary Harsh Vardhan Shringla has clarified that the trilateral security alliance would have little relevance and limited impact on the functioning of the Quad, thereby also showing

---

59. Trevor Findlay, "That Sinking Feeling: The AUKUS Submarines and the Nonproliferation Regime", *AIIA*. https://www.internationalaffairs.org.au/australianoutlook/that-sinking-feeling-the-aukus-submarines-and-the-nonproliferation-regime/

that AUKUS is not necessarily viewed as a negative development.[60] The amount of influence India and Japan hold bilaterally and individually in the Indo-Pacific and the Asian region cannot be matched by Australia or the US alone; Delhi and Tokyo have been consistent long-term contributors to the region with close ties with regional bodies, littoral states and comparatively better ties with China too. Washington must seek to assure India that AUKUS will not limit Washington's focus on the Quad or the India-US partnership and inwardly recalibrate and recalculate the importance it should accord to New Delhi, with a focus on its long-term ambitions in the region.

Concurrently, with North Korea and China keenly watching the future developments vis-à-vis AUKUS, the potential for the two to build their own alliance focussed on aiding nuclear-powered, if not nuclear-weaponised, submarines in an underwater nuclear race, can be a possibility. Russia could emerge as a partner here. Such developments will only seek to further strain the Northeast and East Asian security architecture, and have long-term implications in the form of a rolling arms race. The need for the AUKUS powers to actively subdue international fears regarding their nuclearisation aspects is, hence, vital.

60. "Transcript of Foreign Secretary's Special Briefing on Prime Minister's Visit to USA", Ministry of External Affairs, September 21, 2021. https://mea.gov.in/media-briefings.htm?dtl/34292/transcript+of+foreign+secretarys+special+briefing+on+prime+ministers+visit+to+usa

# 9. Contemporary Missile Developments in United States and Russia post-INF Treaty

*Silky Kaur*

Arms control treaties form the backbone of the nuclear security architecture of the world. The advent of nuclear weapons in 1945 and their subsequent nuclear proliferation led to a series of treaties to contain the spread of nuclear weapons and related technologies. To a great extent, they succeeded in their mission of arresting the spread of nuclear weapons and related technologies, but the second nuclear age is characterised with the undoing of these treaties that came into place after years of negotiations. One such treaty that got undone in 2019 was the Intermediate-Range Nuclear Forces (INF) Treaty.

In this context, firstly, this paper examines the possible reasons for its abandonment. Secondly, it investigates the contemporary missiles development by the United States and Russia post-INF Treaty. Thirdly, it examines the China factor and its shadow on the treaty. Lastly, it seeks to answer what to expect in the future.

This paper argues that in the contemporary age, bilateral treaties are collapsing because of the changed security environment, which has more potential adversaries combined with continuous proliferation of new nuclear missile technologies. To stop the continuing development of nuclear missiles, there is a need to bring more nuclear-armed states to the negotiating table. Hopefully, the ongoing strategic stability dialogues between the US and Russia and the US and China will result in new treaties and agreements to stop the proliferation of nuclear weapons and their delivery systems.

## Intermediate-Range Nuclear Forces Treaty

The INF Treaty was a landmark arms control and disarmament bilateral treaty between the United States and Soviet Union. It was signed on December 8, 1987, by then US President Ronald Reagan and Soviet General Secretary Mikhail Gorbachev. It sought to eliminate their nuclear and conventional ground-launched ballistic and cruise missiles with a range of 500-5,500 km. However, this treaty exempted sea-based and air-launched missiles of these ranges.[1]

This treaty came into force over the public outcry over the deployment of the SS-20 intermediate-range ballistic missile by the Soviet Union and the deployment of Pershing-2 rockets and BGM-109G nuclear-armed Ground-Launched Cruise Missiles (GLCMs) by the US in Europe. These missile deployments in the mid-1970s led to a call for control of intermediate-range missiles. When Mikhail Gorbachev came into power, the INF Treaty negotiations took off. Thus, the INF Treaty was signed on December 8, 1987, and entered into force on June 1, 1988.[2]

By eliminating missiles of the intermediate range category, this treaty became a turning point in the relations between the United States and Soviet Union. Moreover, it also employed extensive on-site inspections for verification. This treaty had succeeded in destroying "a total of 2,692 short-medium and intermediate-range missiles" by June 1, 1991.[3] It also embraced "double-zero options", in which two categories of nuclear weapons were abolished. It was a surprising outcome and was a qualitative change from past treaties, which were focussed on numerical ceilings only.[4]

---

1. US Department of State, archived content, "Treaty Between The United States of America and the Union of Soviet Socialist Republics on the Elimination of Their Intermediate-Range and Shorter-Range Missiles (INF Treaty)". https://2009-2017.state.gov/t/avc/trty/102360.htm#text. Accessed on October 20, 2021.
2. Daryl Kimball and Kingston Reif, "The Intermediate-Range Nuclear Forces (INF) Treaty at a Glance", August 2019. https://www.armscontrol.org/factsheets/INFtreaty. Accessed on October 21, 2021.
3. Ibid.
4. Jeffrey W. Knopf, "Beyond Two-Level Games: Domestic–International Interaction in the Intermediate-Range Nuclear Forces Negotiations", *International Organization*, Vol. 47, No. 6, 1993, p. 600.

In addition, both parties also agreed "to destroy all INF Treaty-related training missiles, rocket stages, launch canisters and launchers".[5]

**Table 1: US and Soviet Missiles Eliminated under INF Treaty**

| United States | Range (kms) | Soviet Union/Russia | Range (kms) |
|---|---|---|---|
| Pershing-II (MGM-31B) (Ballistic Missile) | 1,700 | SS-20 (Ballistic Missile) | 5,000 |
| Pershing 1a (under MGM-31A) (Ballistic Missile) | 740 | SS-4 (Ballistic Missile) | 2,500 |
| BGM-109G Gryphon (Cruise Missile/ GLCM) | 2,500 | SS-5 (Ballistic Missile) | 4,500 |
| | | SS-12 (Ballistic Missile) | 800 |
| | | SS-23 (Ballistic Missile) | 500 |
| | | SSC-X-4 Slingshot (Cruise Missile) | 3,000 |

Source: Compiled by the author from *Arms Control Today*, *The Bulletin of the Atomic Scientists* and US Department of State archived content.

In 2014, the United States first alleged that Russia was violating the INF Treaty by pointing to the development of a GLCM known as the 9M729. Russia consistently denied that it had violated the INF Treaty. However, the US remained unconvinced, and on October 20, 2018, President Trump announced the termination of the INF Treaty. In December 2018, Secretary of State Mike Pompeo stated that Russia was in "material breach" of the INF Treaty. If within 60 days, Russia did not return to full and verifiable compliance, then the US would suspend its obligations, which it did on February 2, 2019, when the 60-day period ended.[6] The next day, Russia also withdrew from the treaty.

5. Kimball and Reif, n. 2.
6. Amy F Woolf, "US Withdrawal from the INF Treaty: What's Next?" Federation of American Scientists, January 2, 2020. https://sgp.fas.org/crs/nuke/IF11051.pdf. Accessed on October 10, 2021. Also see Kimball and Reif, n. 2.

## Missile Developments in Post INF Treaty Environment

With the demise of the INF Treaty, the US and Russia have accelerated their pursuit of the once banned range of missiles. This section examines the missiles they have been testing since abandoning the treaty.

## United States Missile Development Post INF Treaty

The US' departure from the INF Treaty increased the momentum of "development of land-based medium-range cruise, ballistic, and hypersonic systems".[7] The US is modernising its whole strategic triad and also homing in on "developing strategic systems for limited nuclear strikes". These constitute weapons such as the "Trident-2 Submarine-Launched Ballistic Missiles (SLBMs) with low-yield W76-2 warheads, B61-21 variable-yield gravity bombs for heavy bombers and tactical-strike aircraft, long-range stand-off air-launched nuclear cruise missiles, and nuclear sea-based cruise missiles".[8] The US is refocussing on missiles of shorter and intermediate range.

The Trump Administration had asked for a budget of "$100 million in the Fiscal Year (FY) 2020 to develop three new missile systems that would exceed the range limits" of the INF Treaty.[9] It was seeking to test a ground-launched variant of the Tomahawk Submarine-Launched Cruise Missile (SLCM) with a range of 3,000-4,000 km. The budget also had a request to begin the development of mobile, land-based, medium-range missiles capable of attacking "specific threat vulnerabilities in order to penetrate, disintegrate and exploit in the strategic and deep maneuver areas" with a range of 1,000 to 3,000 km.[10] These missiles are expected to fill a gap in the army's development of longer-range missiles and is intended to be fielded by 2023. The US' plans to develop and deploy

---

7. Steven E. Miller and Alexey Arbatov, "Mad Momentum Redux? The Rise and Fall of Nuclear Arms Control". https://www.amacad.org/publication/nuclear-perils-new-era/section/3. Accessed on November 12, 2021.
8. Ibid.
9. Kingston Reif, "Trump Increases Budget for Banned Missiles", Arms Control Association, May 2019. https://www.armscontrol.org/act/2019-05/news/trump-increases-budget-banned-missiles. Accessed on November 12, 2021.
10. Ibid.

intermediate-range missiles have raised concerns among nations. Some, however, have supported the decision. For Japan, the cruise missiles are "exactly the kind of defensive capability that countries such as Japan will want and will need for the future".[11]

Since November 2021, the US Army has also been reactivating its 56th Artillery Command in Germany. It will increase readiness and multi-national interoperability by integrating joint and multi-national forces in theatre operations and exercises. This unit was active between 1963 and 1991, and commanded battalions armed with the Pershing I and Pershing II nuclear-armed ballistic missiles. This artillery group was deactivated after the successful implementation of the INF Treaty but the collapse of the treaty in 2019 has led to its reactivation. Now, in the coming years, it will become a hub for artillery operations across Europe, with deployments of new ground-based hypersonic weapons and other longer-range missiles.[12]

### *New Ground-Launched Cruise Missile*

After withdrawing from the INF Treaty on August 2, 2019, the US tested the new GLCM, a variant of the Tomahawk SLCM on August 18, 2019. This missile test was in response to Russia's treaty violations and its fielding of similar ground-launched missiles. It was launched off the coast of California on San Nicolas Island. In a statement, the department said the "test missile exited its ground mobile launcher and accurately impacted its target after more than 500 kilometers of

11. Kingston Reif and Shannon Bugos, "US Aims to Add INF-Range Missile", October 2020. https://www.armscontrol.org/act/2020-10/news/us-aims-add-inf-range-missiles. Accessed on November 12, 2021.
12. Caleb Larson, "The United States Army is Reactivating a Cold War- era Artillery Command", *The National Interest*, November 9, 2021. https://nationalinterest.org/blog/buzz/united-states-army-reactivating-cold-war-era-artillery-command-195918. Accessed on November 30, 2021. Also see Joseph Trevithick "Army Revives Cold War Nuclear Missile Unit to Deploy New Long-Range Weapons in Europe", November 8, 2021. https://www.thedrive.com/the-war-zone/43051/army-revives-cold-war-nuclear-missile-unit-to-deploy-new-long-range-weapons-in-europe. Accessed on November 30, 2021.

flight."[13] The MK-41 launcher was used, and it was affixed to a mobile trailer. The US has "dozens of Mark 41 launchers in Romania and Poland as part of a regional ballistic missile defence system".[14] Russia has objected that Mark 41 launchers themselves were in INF Treaty violations.

In background history, the BGM-109 Gryphon GLCM was a land-based, ground-launched nuclear armed variant of the Tomahawk Land Attack Missile (TLAM), with a W84 nuclear warhead; and was stationed in Europe in the 1980s until the INF Treaty ratification in 1987. This GLCM deployment in Europe led to a huge public outcry because the deployment of GLCMs along with the Pershing IIs was highly destabilising.[15]

Thus, the BGM-109G was withdrawn from service in 1991 to comply with the INF Treaty. It was a counter to the Intermediate-Range Ballistic Missiles (IRBMs) and Medium Range Ballistic Missiles (MRBMs) of the Soviet Union such as the SS-20 Saber. It is generally stated that the GLCM and Pershing II were the main reasons for the Soviet Union to come to terms with INF Treaty. GLCM is used as a generic name for any ground-launched cruise missile and the US deployed only one modern cruise missile in the tactical role with a nuclear warhead, therefore, the GLCM name stuck. It had a range of 2,000 km.[16]

---

13. Kingston Reif, "Treaty Withdrawal Accelerates Missile Debate", September 2019. https://www.armscontrol.org/act/2019-09/news/treaty-withdrawal-accelerates-missile-debate. Accessed on November 5, 2021.
14. Kyle Mizokami, "After Leaving Treaty, US Tests Ground Launched Cruise Missile" August 20, 2019. https://www.popularmechanics.com/military/weapons/a28750743/ground-launch-missile/. Accessed on November 5, 2021. Also see Shannon Bugos, "US Completes INF Treaty Withdrawal", September 2019. https://www.armscontrol.org/act/2019-09/news/us-completes-inf-treaty-withdrawal. Accessed on November 5, 2021.
15. Ibid.
16. Marcus Weisgerber, "Pentagon Confirms It's Developing Nuclear Cruise Missile to Counter a Similar Russian One", February 2, 2018. https://www.defenseone.com/threats/2018/02/pentagon-nuclear-cruise-missile-russia/145689/. Accessed on November 14, 2021.

Though they were withdrawn from service after the INF Treaty, since 2018, the US Army has been continuously working on new ground-launched, intermediate-range cruise missiles to counter a similar Russian one.[17] Along with the ongoing tests, the 2021 fiscal year budget request included $125 million for purchasing 48 Tomahawk missiles to "assess the feasibility and utility of firing the missile from a ground launcher".[18] The US Army is planning to field this GLCM prototype with a range between 500 and 2,000 km by 2023. The new missiles that will be developed would join other ground-launched missiles with a range formerly prohibited by the treaty along with the precision strike missile and the long-range hypersonic weapon.[19]

### *Ground-Launched Ballistic Missile*

On December 12, 2019, the US conducted a second test of a missile formerly banned by the defunct INF Treaty; it was described as a "prototype, conventionally configured, ground-launched ballistic missile from Vandenberg Air Force Base, California".[20] The missile flew for more than 500 km. This range was prohibited in the INF Treaty. Some have speculated that it was a "Castor IVB rocket motor which has been used in the past for military space launches and in target vehicles for the Missile Defence Agency".[21] It was a follow-up to the August 18, 2019, test of the intermediate range GLCM from a Mark 41 vertical launch system canister. On December 12, 2021, Defence Secretary Mark Esper stated, "Once we develop intermediate-range missiles, and if my commanders require them, then we will work closely and consult closely with our allies in Europe, Asia and elsewhere with regard to any possible deployments".[22]

17. Ibid.
18. Reif and Bugos, n. 11.
19. Ibid.
20. Kingston Reif and Shannon Bugos, "US Tests Second Medium-Range Missile", January/February 2020. https://www.armscontrol.org/act/2020-01/news/us-tests-second-medium-range-missile. Accessed on November 14, 2021.
21. Ibid.
22. Ibid.

### *Precision Strike Missiles*

Precision Strike Missiles (PrSMs) are the newest post-INF Treaty weapons. These are the first new class of a previously prohibited range of missiles.[23] On October 13, 2021, Lockheed Martin's PrSM completed its "longest flight to date, exceeding the maximum threshold, with the US Army, at Vandenberg Space Force Base (VSFB), California".[24] It is noted that this was the fifth consecutive successful flight test for the missile. Moreover, Lockheed Martin got the contract for $62 million on September 30, 2021, for the missile's engineering and manufacturing development phase, and it was intended to be fielded by 2023. It would replace the aging tactical missile system and play a significant role in developing the capability for deep-strike necessary to counter adversaries. The original goal of these missiles was to have the capability of only 499 km, but the INF Treaty's demise led to the development of a range much more than the 499 km limit.[25]

On December 13, 2021, the Russian News Agency TASS reported that the Russian Army would soon conduct drills over its western border using the Iskander operational and tactical missile systems in response to the planned deployment of US medium range missiles in Europe. The Russian Foreign Ministry stated that reactivation of the artillery command in Germany which operated the nuclear capable Pershing missiles during the Cold War, with the planned deployment of the new Precision Strike Missile (PrSM) in place of the Pershing with a range of 500-800 km, is a cause of major concern. Though NATO has claimed

23. Caleb Larson "Lockheed Martin Announces Army's Precision Strike Missile Completes Test", October 16, 2021. https://nationalinterest.org/blog/buzz/lockheed-martin-announces-armys-precision-strike-missile-completes-test-195132. Accessed on December 16, 2021.
24. Lockheed Martin, "US Army Breaks New Ground as Lockheed Martin's PrSM Completes Record Setting Flight", October 14, 2021. https://news.lockheedmartin.com/u-s-army-breaks-new-ground-as-lockheed-martins-PrSM-completes-record-setting-flight. Accessed on December 16, 2021.
25. Jen Judson, "US Army's Precision Strike Missile Gets Green Light for Development, Readies for Big Test", October 11, 2021. https://www.defensenews.com/land/2021/10/11/us-armys-precision-strike-missile-gets-green-light-for-development-readies-for-big-test/. Accessed on November 15, 2021.

that these missiles will not have nuclear warheads, Russia's deputy foreign minister highlighted the lack of trust and said that "Russia cannot control the missiles the US will deploy in Europe and has no guarantees they won't carry nuclear weapons". In response to this, Russia may deploy "medium range-missiles with nuclear warheads on its western borders and possibly in Belarus".[26]

### *US Hypersonic Missile System*

Though the US hypersonic missile system has been declared as conventional, it interacts with Russia's and China's nuclear capability. The presence of an advanced hypersonic system, even if it carries conventional warheads poses a challenge to other nations and has implications for nuclear capabilities. The US is currently developing a range of advanced hypersonic systems. It has also recently given a contract to "Lockheed Martin for the development of two systems: the Hypersonic Conventional Strike Weapon ($928 million) and AGM-183A Air-Launched Rapid Response Weapon ($480 million)".[27]

In May 2021, the Biden Administration released its 2022 fiscal year budget request, in which it asked for "$3.8 billion for all hypersonic-related work". It is focussing on the development of "offensive hypersonic weapons programs as a part of a continued emphasis on strengthening conventional deterrence against China and Russia".[28] In October 2021,

26. TASS, "Press Review: Putin, Xi to Hold Video Conference Call and Kiev Readies Bomb Shelters" December 15, 2021. https://tass.com/pressreview/1375717, accessed on December 20, 2021. Also see Alexander Marrow and Mark Trevelyan, "Russia Says it May be Forced to Deploy Mid-Range Nuclear Missiles in Europe", December 13, 2021. https://www.reuters.com/world/russia-says-lack-nato-security-guarantees-would-lead-confrontation-ria-2021-12-13/. Accessed on December 20, 2021.
27. Partyard Military Division, "Hypersonic Missiles: What are They and Can They be Stopped?" May 10, 2019. https://partyardmilitary.com/hypersonic-missiles-what-are-they-and-can-they-be-stopped/. Accessed on November 20, 2021.
28. Shannon Bugos and Kingston Reif, "Understanding Hypersonic Weapons: Managing the Allure and the Risks", *Arms Control Association Report*, September 2021. https://www.armscontrol.org/sites/default/files/files/Reports/ACA_Report_HypersonicWeapons_2021.pdf. Accessed on November 20, 2021.

US forces conducted three tests of hypersonic missile component prototypes.[29] The US Department of Defence is funding more than "eight prototype hypersonic weapons programs with the aim of fielding an initial capability of some of those by 2022". It is speculated that "while Washington is pursuing conventional hypersonic weapons at this time, Beijing and Moscow appear to be seeking not only conventional but also nuclear or dual-capable hypersonic capabilities".[30]

**Fig 1:**

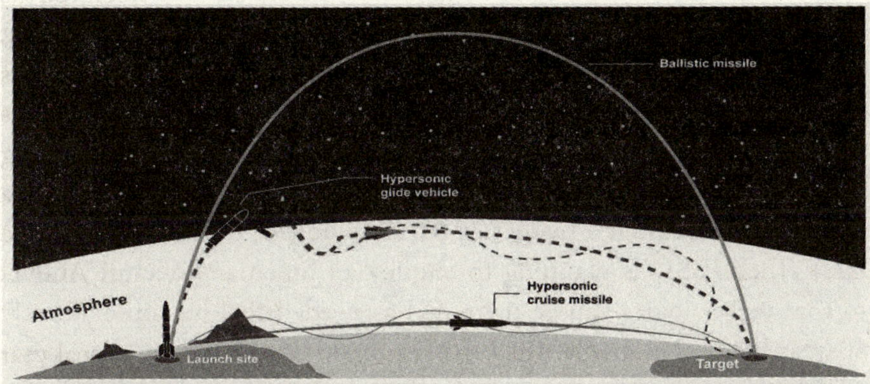


**Ballistic missiles follow a largely predictable, arched trajectory, flying high above the atmosphere before plummeting back towards Earth. Hypersonic glide vehicles and cruise missiles fly at lower altitudes within the atmosphere and follow different flight paths.** (Illustration by United States Government Accountability Office)

Source: Arms Control Association[31]

Hypersonic weapons are a challenge to existing air and missile defence systems. Currently, there is no defence against hypersonic weapons. The defence against hypersonic weapons will require a new intercept system; also Hypersonic Glide Vehicles (HGVs) pose problems "for broad-area midcourse missile defenses which are the backbone of the US missile defence architecture".[32] The US Army is also requesting a $801 million

29. Al Jazeera, "New Space Race? US Tests Components for Hypersonic Weapons," October 22, 2021. https://www.aljazeera.com/news/2021/10/22/new-space-race-us-tests-components-for-hypersonic-weapons. Accessed on November 20, 2021.
30. Bugos and Reif, n. 28.
31. Ibid.
32. Ibid.

budget so that its long-range hypersonic weapons programme can pair the common glide vehicle with the navy's booster system. It will provide the army with a "prototype strategic attack weapon system" and is intended to be fielded in 2023".[33]

It is to be noted that short and medium range systems are considered as non-strategic nuclear weapons. In American parlance, they are known as "battlefield, tactical and theater nuclear weapons" and "long-range missiles and heavy bombers are known as strategic nuclear delivery vehicles".[34] In the realm of strategic nuclear delivery vehicles, the US Air Force has modernised the Minuteman missiles by "replacing and upgrading their rocket motors and guidance systems and other components, so that they can remain in the force through 2030".[35] It has also begun a programme to replace these Minuteman missiles with a new Ground-Based Strategic Deterrent (GBSD) beginning around 2030. For this, it had requested a budget of $1.52 billion in Financial Year (FY) 2021. The GBSD is planning to employ an "open architecture" model so that technologies can be upgraded as needed for the approximately 60-year life of the missile. The GBSD programme would be "more cost-effective than the life-extension program because it would provide the US with modern missiles with greater capabilities and lower maintenance costs. Also, there are plans to replace the Air-Launched Cruise Missile (ALCM) with the new Long-Range Stand-Off (LRSO) Missile".[36]

The US Navy is also extending the life of the Trident II (D-5) SLBMs. And new Columbia class submarines are going to replace the currently deployed Ohio class submarines by 2031. The US Air Force, along with procuring the LRSO cruise missiles is also working on extending the

33. Kelley M. Sayler, "Hypersonic Weapons: Background and Issues for Congress" Congressional Research Service, October 19, 2021. https://sgp.fas.org/crs/weapons/R45811.pdf. Accessed on December 15, 2021.
34. Amy F. Woolf, "US Strategic Nuclear Forces: Background, Developments, and Issues", Congressional Research Service, December 14, 2021. https://sgp.fas.org/crs/nuke/RL33640.pdf. Accessed on December 20, 2021.
35. Ibid.
36. Ibid.

life of the B61 bomb which was carried on B-2 bombers, also on fighter aircraft and the W80 warhead for cruise missiles.[37]

*Basing Dilemmas of Intermediate-Range Missiles*

Regarding basing of these missiles, several countries in Europe have opined that any deployment of the INF Treaty range missiles would need to be approved by all NATO members. Russia and China also have criticised such possible deployments. Vladimir Putin said that if the US develops and deploys intermediate range missiles, Russia will also engage in a full-scale effort to develop similar missiles. President Putin also said that the US test of the ground-launched Tomahawk missile shows that a new threat is appearing, and Russia should prepare itself to respond to such threats.[38] As stated earlier, Russia has continuously objected to the deployment of the MK-41 which can launch not only anti-ballistic missiles but also combat cruise missiles.[39]Meanwhile, Russia tabled two draft treaties on December 17, 2021, namely the Treaty between the US and Russian Federation on Security Guarantees and the Agreement on Measures to Ensure the Security of the Russian Federation and Member States of NATO.[40] In these treaties, Russia has called for restraints on INF range deployments in Europe and rollback of military capabilities in new NATO states to the 1997 deployments.[41]

China also stated that if the US engages in deployments of intermediate-range missiles in the East Asian part of the world, on its doorsteps, China will take counter-measures.[42] For instance, the

37. Ibid.
38. Reif, n. 13.
39. Ibid.
40. Treaty between the United States of America and the Russian Federation on Security Guarantees, December 17, 2021. https://mid.ru/ru/foreign_policy/rso/nato/1790818/?lang=en. Accessed on December 21, 2021.
41. Patricia Lewis, "Russian Treaty Proposals Hark Back to Post-Cold War Era" December 19, 2021. https://www.chathamhouse.org/2021/12/russian-treaty-proposals-hark-back-post-cold-war-era. Accessed on December 22, 2021.
42. Reif, n. 13.

deployment of the Terminal High Altitude Area Defence (THAAD) system in South Korea by the US in 2017, led to strong opposition by China. China believes that though it is said that THAAD is for defence against North Korean ballistic missiles, in reality, it has been deployed to weaken China's nuclear-deterrent forces and is a threat to both China and Russia. Also, THAAD deployment is seen as a kind of betrayal by South Korea. It has become a major irritant for China in the Northeast Asian region.[43] China fears that this fielding of THAAD in Korea will "reinforce and reshape US alliances in Northeast Asia, both by tightening the alliance with South Korea and by fostering a trilateral US-Republic of Korea (ROK)-Japan security relationship". It may also spark an arms race on the Korean Peninsula.[44]

## Russia's Nuclear Missile Development post-INF Treaty

After withdrawing from the INF Treaty, Russian Defence Minister Sergei Shoigu said that in response to the end of the INF Treaty, Russia should focus on developing new hypersonic and cruise ground-based missiles.[45] Russia is working on a new generation of "nuclear and dual-purpose weapon systems" which constitute weapons such as the "Avangard strategic nuclear boost-glide hypersonic system; Poseidon long-range, high-speed, nuclear-propelled and nuclear-armed heavy torpedoes; Burevestnik nuclear-powered intercontinental nuclear cruise missiles; Kinzhal air-launched hypersonic middle-range missiles; and a number of other sub-strategic nuclear and dual-purpose

43. Michael D Swaine, "Chinese Views on South Korea's Deployment of THAAD", Carnegie Endowment. https://carnegieendowment.org/files/CLM52MS.pdf. Accessed on December 17, 2021.
44. Robert C. Watts IV, "Rockets' Red Glare"- Why Does China Oppose THAAD in South Korea and What Does it Mean for US Policy", *Naval War College Review*, Vol 71, No. 2, 2018. https://digital-commons.usnwc.edu/cgi/viewcontent.cgi?article=1736&context=nwc-review. Accessed on December 17, 2021.
45. David Reid, "Russia Says it Must Develop New Nuclear Warhead-Carrying Missiles by 2021", cnbc.com, February 5, 2019. https://www.cnbc.com/2019/02/05/russia-says-it-must-develop-new-nuclear-missiles-after-inf-end.html. Accessed on September 1, 2021.

systems".[46] With the withdrawal from the INF Treaty, some of these missiles, which are of intermediate range such as the "intermediate-range land-based Kalibr-type cruise missiles and hypersonic missiles", are being developed and deployed. The US government has alleged that Russia has already deployed a "ground-launched cruise missile similar to the Kalibr 3M14".[47]

Recently, the US Defence Department also stated that Russia is continuously modernising, "an active stockpile of up to 2,000 non-strategic nuclear weapons, including those employable by ships, planes, and ground forces".[48] Some of these weapons constitute "air-to-surface missiles, short range ballistic missiles, gravity bombs, and depth charges for medium-range bombers, tactical bombers, and naval aviation, as well as anti-ship, anti-submarine, and anti-aircraft missiles and torpedoes for surface ships and submarines, and nuclear ground-launched cruise missile".[49]

### *9M729/SSC-8*

Russia has developed and deployed this missile in violation of the INF Treaty. This missile is reported to support both nuclear and conventional weapons. It has a range of 2,500 km. As stated earlier, its development triggered the 2019 US withdrawal from the 1987 INF Treaty. It is reported to have been in service since 2017.[50] Former National Intelligence Director Dan Coats stated that Russia tested the 9M729 from a fixed launcher under prohibited ranges, then again tested it from a mobile launcher to permitted ranges. The first two battalions of these weapons were deployed in late 2017, and later Russia "deployed

46. Miller and Arbatov, n. 7.
47. Ibid.
48. Hans M. Kristensen and Matt Korda, "Russian Nuclear Weapons, 2021", *The Bulletin of the Atomic Scientists,* Vol. 77, No. 2, March 18, 2021. https://www.tandfonline.com/doi/full/10.1080/00963402.2021.1885869. Accessed on November 12, 2021.
49. Ibid.
50. Missile Defence Project, "9M729 (SSC-8)," Centre for Strategic and International Studies, October 23, 2018, last modified July 31, 2021. https://missilethreat.csis.org/missile/ssc-8-novator-9m729/.

four battalions in the Western, Southern, Central, and Eastern Military Districts with nearly 100 missiles".[51]

*Burevestnik Nuclear Propelled Cruise Missile/RS-SSC-X-09*

Also known as "Skyfall", this is Russia's nuclear powered cruise missile. It is characterised as 'developmental' and of unlimited range. It has the ability to evade missile defences because it flies under the ballistic weapon shield. It can penetrate any interceptor-based missile defence system.[52] The earlier missile tests faced serious setbacks, and in 2017, a failed test resulted in the missile being lost at sea. In 2019, an explosion killed five scientists and two soldiers at Nenoksa.[53] But, development has continued.

*Avangard Boost-Glide Hypersonic Project*

The Avangard is a nuclear-capable, hypersonic boost-glide vehicle. During a speech in March 2018, President Putin described it as one of the six "next generation" weapons. It has a range of over 6,000 km and weighs approximately 2,000 kg. It can carry a nuclear or conventional payload. According to a TASS report, the "HGV's nuclear warhead is more than 2 megatons in TNT equivalent".[54] As a boost-glide weapon, the Avangard is carried to its apogee by a ballistic missile. Currently, the carrier is the SS-19 "Stiletto" also known as the "UR-100NUTTH, but will later be replaced by the R-28 Sarmat."[55]

The HGV can manoeuvre and maintain speeds of up to Mach 20. This manoeuverability makes the Avangard's "trajectory unpredictable,

51. Kristensen and Korda, n. 48.
52. Douglas Barrie and Henry Boyd, "Burevestnik: US Intelligence and Russia's 'Unique' Cruise Missile", The International Institute for Strategic Studies, February 5, 2021. https://www.iiss.org/blogs/military-balance/2021/02/burevestnik-russia-cruise-missile. Accessed on December 20, 2021.
53. Kristensen and Korda, n. 48.
54. Missile Defence Project, "Avangard," Missile Threat, Centre for Strategic and International Studies, July 31, 2021. https://missilethreat.csis.org/missile/avangard/. Accessed on November 12, 2021.
55. Ibid.

complicating intercept attempts after its boost phase". The HGV does not "employ a propulsion system, it relies on gravity and its aerodynamic features to maintain speeds and altitude".[56]

President Putin in March 2018 stated that the Avangard HGV had already entered serial production. In December 2020, Defence Minister Sergey Shoigu said that in the Strategic Missile Force, 31 launchers with the Yars and Avangard ICBMs will undertake combat duty.[57] By the end of 2027, the second regiment of six missiles will reportedly be added.[58]

### *Tsirkon/ Zircon/ 3M22 Tsirkon/SS-N-33*

This is a sea-launched anti-ship hypersonic cruise missile of Russia; it has a range of 500-1,000 km.[59] It is a strategic missile and has an extremely fast speed of Mach 8 which can be accelerated to Mach 9. It is considered as the fastest missile in the world. Due to its speed, defence against this missile is very difficult. Another feature of the missile is its "plasma cloud": during flight, "the missile is completely covered by a plasma cloud that absorbs any rays of radio frequencies and makes the missile invisible to radars". This makes the missile undetectable on its way to the target. The latest test of the Zircon was conducted on November 29, 2021, from the waters of the White Sea.[60]

### *Hypersonic Cruise Missile Kinzhal/ Kh-47 M2 Kinzhal*

This is an air-launched hypersonic missile also known as the "dagger". It has a range of 2,000 km. It can carry a nuclear or conventional

56. Ibid.
57. Ibid.
58. Kristensen and Korda, n. 48.
59. Naval Technology, "Russia Conducts Another Test Launch of Zircon Hypersonic Missile", November 29, 2021. https://www.naval-technology.com/news/russia-test-zircon-hypersonic-missile/. Accessed on November 30, 2021.
60. Missile Defence Advocacy Alliance, "3M22 Zircon". https://missiledefenseadvocacy.org/missile-threat-and-proliferation/todays-missile-threat/russia/3m22-zircon/. Accessed on December 20, 2021.

warhead. The missile is said to be similar to the ground-launched SS-26 short-range missile. Its range is up to 2,000 km. It can be launched from the "center-pylon of specially modified MiG-31 K Foxhound air interceptors". This missile can be used against both land and sea targets. It has also been reportedly deployed on "experimental combat duty in the Southern Military District since December 2017". The Kinzhal missile was "publicly demonstrated for the first time in an air show in August 2019".[61]

## The China Factor in the Abandonment of the INF Treaty

The INF was a bilateral treaty which left China free to develop INF range missiles, which it did. Several US strategists believed that the INF Treaty placed the US at a disadvantage in the context of challenges faced from China. China has reportedly deployed "thousands of land-based intermediate-range ballistic and cruise missiles that threaten US forces and allies in Asia". So along with Russia, China was also a factor behind the US' exit from the INF Treaty.[62] There were concerns that China would continue to develop nuclear weapons while the US would be constrained from doing so.

During the three decades that the INF Treaty was in place, China worked on its intermediate missile capabilities to develop one of the biggest arsenals of INF range missiles. China presently has 1,200 of these DF-11s, DF-15s and DF-16s. Secondly, it has 250 DF-21 missiles with a range of more than 2,150 km. Thirdly, these are supported by approximately 300 DH-10 cruise missile which have a range of 2,000 km. Fourthly, China has 30-36 DF-26 ballistic missiles with conventional and nuclear warheads and a range of 4,000 km.[63]

61. Kristensen and Korda, n. 48.
62. Woolf, n. 6.
63. Marcin Andrzej Piotrowski, "China's Approach to the INF Treaty and the Development of Intermediate Range Missiles". https://pism.pl/publications/China_s_Approach_to_the_INF_Treaty_and_the_Development_of_Intermediate_Range_Missiles. Accessed on December 20, 2021.

**Table 2: China's Intermediate Range Nuclear Missiles**

| Missiles (Nuclear) | Deployed | Range (kms) |
|---|---|---|
| DF-21 | 1991 | 2,150 |
| DF-21 A | 1996 | 2,150 |
| DF-21 C | 2006 | 2,150 |
| DF-16 | 2011 | 1,000 |
| DF-17 | 2019 | 2,500 |
| DF-26 | 2016 | 4,000 |
| DF-11A | 1992 | 600 |
| DF-15B | 2009 | 725 |
| DH-10 | 2006 | 2,000 |

Source: Compiled by the author from IISS[64], CSIS[65] and other sources.

These missiles below the range of 5,500 km constitute around 93 per cent of the land-based ballistic and cruise missiles of China. They are nearly 1,250 in number and are a result of a multi-decade focus on the development of these systems. They are strategically useful for China in enhancing its "coercive and kinetic capabilities to confront, deter and, if necessary, defeat the military challenges on their doorstep".[66]

China has also developed hypersonic weapons. The Chinese Academy of Sciences launched the "shock tunnel reproducing hypersonic flight conditions" in 2008. This JF-12 hypersonic wind tunnel is being used for the development of the Starry Sky HGV. The Starry Sky-2 or Xing Kong-2 can "carry nuclear warheads and can travel at six times the speed

64. Bates Gill, "Exploring post-INF Arms Control in the Asia-Pacific: China's Role in the Challenges Ahead", Research Papers, International Institute for Strategic Studies, June 29, 2021. https://www.iiss.org/blogs/research-paper/2021/06/post-inf-arms-control-asia-pacific-china. Accessed on December 18, 2021.
65. Missile Defence Project, "Missiles of China," Centre for Strategic and International Studies, April 12, 2021. https://missilethreat.csis.org/country/china/. Accessed on December 18, 2021.
66. Gill, n. 64.

of sound".[67] It is a nuclear capable hypersonic vehicle prototype. It is also known as "waverider" for its capability to derive lift from the shockwaves generated by its own hypersonic flight.[68]

Moreover, China's Dongfeng-17 is considered as a dual-capable (conventional and nuclear) medium-range ballistic missile system featuring a hypersonic glide vehicle with a range of 1,800-2,500 km and a speed of Mach 5 and 10.[69] During the 2019 military parade, China "displayed a ballistic missile designed specifically to carry a hypersonic glide vehicle, the DF-17". China's motivation to develop hypersonic weapons lies mostly in defeating the "US offensive weapons in the event of a major conflict in Asia".[70]

China perceives the US withdrawal from the INF Treaty and the plans for future deployment of ground-based missiles to Asia as part of the strategy of the US to contain China. This can lead to a missile race in Asia. For a long time, the INF Treaty acted as a "security guarantee for China" because the limitations the US and Russia due to the treaty were beneficial for China. Now in this era of open confrontation between China and the US, the latter needed to free itself from the restrictions posed by the INF Treaty. The INF Treaty restricted the US from deploying ground-based intermediate-range missiles in Asia. China perceives the INF Treaty's demise as a US strategy to contain China by deploying ground-based missiles in Asia.[71]

## What to Expect in the Future?

Since the past few years, though most of the NATO members have echoed the US sentiments regarding Russia's breach of the INF

67. Holmes Liao, "China's Development of Hypersonic Missiles and Thought on Hypersonic Defence", *China Brief* Vol. 21; Issue:19, October 8, 2021. https://jamestown.org/program/chinas-development-of-hypersonic-missiles-and-thought-on-hypersonic-defence/. Accessed on November 26, 2021.
68. Bugos and Reif, n. 28.
69. Ibid.
70. Andrey Baklitskiy, "What the End of the INF Treaty Means for China", Carnegie Moscow Centre, https://carnegiemoscow.org/commentary/80462. Accessed on October 10, 2021.
71. Ibid.

Treaty some members have also expressed doubts on the claim that Russia has violated the INF Treaty.[72] The US withdrawal also left many members shocked. Foreign Minister of Germany Heiko Maas stated that the "INF Treaty had been an important pillar of the European security architecture for over 30 years" and the US should consider the "consequences both for Europe and for the future disarmament efforts".[73] Japan also termed the demise of the INF Treaty as undesirable because this would accelerate the arms race with Russia and China and prove to be a stumbling block in progress towards North Korea's denuclearisation.[74] However, on the other side, for Japan, the post-INF environment in Asia is also an opportunity for increasing deterrence to the Pacific region by the deployment of new American capabilities. Beijing's continuous progress in missile development in the INF range was leading to a missile gap in the East Asian security environment. For Japan, the US withdrawal from the INF Treaty addresses the missile gap with China.[75]

Changes in the international security environment also led to the INF Treaty's demise. When the treaty was signed in 1987, the Soviet Union was in existence. Superpower tension was at its peak. Both superpowers were installing medium-range missiles across Europe. The danger was centred in Europe. Now in the post-Cold War world, the security environment has changed. Nuclear weapons and their delivery vehicles are proliferating in other parts of the world. The biggest challenge today from the US' perspective is not Russia but China: in view of the speed at which China is accelerating its military capabilities, the US is also seeking to have a substantial land-based missile presence in Asia.[76]

72. Woolf, n. 6.
73. Ibid.
74. Ibid.
75. Ankit Panda, "The US and Japan After the INF Treaty". https://spfusa.org/wp-content/uploads/2020/02/The-US-and-Japan-After-the-INF-Treaty_FINAL.pdf. Accessed on November 8, 2021.
76. Daryl Kimball and Kingston Reif, "The Intermediate-Range Nuclear Forces (INF) Treaty at a Glance", August 2019. https://www.armscontrol.org/factsheets/INFtreaty. Accessed on October 21, 2021.

Basically, the second nuclear age has three main characteristics: firstly, there is a diminishing role of treaties and a near-collapse of the arms control regime between the US and Russia, as the only remaining treaty is the New Strategic Arms Reduction Treaty (START) that has been extended till 2026. Secondly, there is an immense change in nuclear military technologies. Thirdly, the nuclear-armed states are now facing two or more potential adversaries, multiple threats, and security trilemmas.[77] The existing arms control treaties are not able to arrest the relentless development of new types of nuclear weapons delivery missiles. For instance, the hypersonic weapons can prompt an escalation because of target and warhead ambiguity. Also, they can escalate a conflict instantly due to their high speeds and the reduction in response time.

Given the complexities involved in restraining the spread of nuclear missiles in the future, there will be a continuous arms race unless the Strategic Stability Dialogue can bring some arms control. The Strategic Stability Dialogue initiated by the Biden Administration with Russia may succeed in charting out a path for dealing with proliferation and redeployment of new missile capabilities.[78] Besides Russia, however, the current nuclear scenario also requires bringing China into the arms control regime. The US wants to include China in the arms control talks. China has largely declined the proposals to join the trilateral talks on nuclear arms control. Meanwhile, Russia wants to bring France and the United Kingdom into the arms control negotiations.[79] It remains to

---

77. Augusto C. Dall Agnol and Marco Cepik, "The Demise of the INF Treaty: A Path dependence analysis", SciELO Brazil, 2021. https://doi.org/10.1590/0034-7329202100202. Accessed on November 15, 2021.
78. The White House, *U.S.-Russia Presidential Joint Statement on Strategic Stability,* Press release, June 16, 2021. https://www.whitehouse.gov/briefing-room/statements-releases/2021/06/16/u-s-russia-presidential-joint-statement-on-strategic-stability/. Accessed on September 25, 2021.
79. Reuters, "Russia Touts Britain, France for Wider Nuclear Talks", July 29, 2021. https://www.reuters.com/world/china/russia-says-us-wants-china-included-arms-control-talks-ifax-2021-07-29/. Accessed on September 30, 2021.

be seen if future rounds of the Strategic Stability Dialogue can bring some measures and controls for restraining the spread of intermediate range nuclear missiles. Till this happens, it is likely that there will be an offence-defence spiral of growth of missiles in intermediate ranges.

# 10. Nuclear Missile Development Trends in Southern Asia: An Analysis from 2011 to 2021

*Nasima Khatoon*

## Introduction

Over the last decade, China, Pakistan and India have been steadily developing and modernising their nuclear missiles and armed forces to address their respective security concerns and deterrent requirements. Delivery systems are critical to credible nuclear deterrence. Acquisition of nuclear weapons without an effective delivery system is of little value. Therein lies the importance of delivery systems like ballistic and cruise missiles, aircraft and sea-based delivery platforms. These delivery platforms are perceived as necessary to ensure range, reliability and survivability. The trends in nuclear missile development in the region over the previous ten years are examined in this chapter.

## Major Missile Trends in China, India and Pakistan

China initiated reform in the People's Liberation Army (PLA) primarily from 2013 onwards to "fully transform the people's armed forces into world-class forces by the mid-21st century"[1]. The modernisation aims of the PLA include improvements in strategic capability, military organisational structure, weaponry and comprehensive reform of

1. "China's National Defence in the New Era", 2019 Defence White Paper, The State Council Information Office of the People's Republic of China, July 2019. Accessed on August 30, 2020.

national defence to transform it into a world-class force in the coming decades.

A significant part of this transformation is China's nuclear missile force that has undergone numerous technological advancements and organisational and operational changes over the past decade. For China, the People's Liberation Army Rocket Force (PLARF) is the primary military organisation responsible for operating and maintaining its conventional and nuclear missile forces. The PLARF was elevated to the level of a full Service alongside the army, navy and air force in 2015.

In the past decade, China has worked on a range of technologies to alleviate its threat perceptions. The most significant technological developments include the Multiple Independently Targetable Reentry Vehicles (MIRVs), Manoeuvrable Reentry Vehicles (MaRVs), the successful development of Submarine-Launched Ballistic Missile (SLBM) capability, hypersonic weapon capability and solid-fuelled Intercontinental Ballistic Missile (ICBM) capability. Apart from this, China is possibly also developing an air-launched dual capable ballistic missile.

In the case of Pakistan, some of the significant technological developments that can be observed are the deployment of Tactical Nuclear Weapons (TNWs), development of MIRV technology, improvements in the existing and new nuclear capable cruise missiles and Submarine-Launched Cruise Missiles (SLCMs) in order to complete the nuclear triad. While Pakistan is yet to completely develop and operationalise most of these technologies, the pace of development has increased substantially in the recent years through missile tests and assistance by China.

At the same time, significant developments in India include enhancement of its sea-based nuclear delivery platforms, long range ballistic missile capability, development of mobility and accuracy of ballistic missiles and possible development of MIRV capability, and some initial testing of hypersonic technology. These developments are primarily influenced by its threat perceptions and the need to strengthen its operational capabilities. In order to achieve these objectives, India has invested significantly in its land, air and sea-based nuclear deterrence while keeping its goals defensive and limited.

Having specified the major trends of nuclear missile development in the region, the following section analyses the above-mentioned trends.

## Increase in the Types of Missiles and Testing

A recent estimate of nuclear weapons stockpiles shows that as of November 2021, China maintained a stockpile of approximately 350 nuclear warheads, compared to an estimated stockpile of 240 nuclear warheads in 2011.[2] An interesting trend can be observed regarding the numbers of land-based and sea-based ballistic missiles. According to the estimate of *The Bulletin of the Atomic Scientists,* presently, China has approximately 280 nuclear capable land-based ballistic missiles and 72 sea-based ballistic missiles.[3] The US Department of Defence (DoD) report, 2021, estimates that of these missiles, around 150 are ICBMs. With the operationalisation of the ICBMs DF-5C and DF-41, this number is likely to grow further. Whereas the estimated number of missiles (land-based ballistic missiles) till 2011 was nearly 140. Considering the fact that some of these missiles are dual capable in nature, this growth impacts the overall capability of the missile force.

The 2021 US DoD report estimated that China has approximately 100 ICBM launchers deployed at missile bases, having increased from around 60 ICBM launchers in 2010.[4] This excludes the number of DF-41 ICBM launchers. Besides, road mobile ICBM launchers, China is also building a large number of missile silos. Recent evidence indicates

2. Hans M. Kristensen and Matt Korda, "Nuclear Notebook: Chinese Nuclear Forces", *The Bulletin of the Atomic Scientists*, November 15, 2021. https://thebulletin.org/premium/2021-11/nuclear-notebook-chinese-nuclear-forces-2021/. Accessed on December 10, 2021.
3. Ibid.
4. Office of the US Secretary of Defence, "Military and Security Developments Involving the People's Republic of China 2021", Annual Report to Congress. https://media.defence.gov/2021/Nov/03/2002885874/-1/-1/0/2021-CMPR-FINAL.PDF and US DoD report 2010 at https://dod.defence.gov/Portals/1/Documents/pubs/2010_CMPR_Final.pdf.

that China is constructing more than 250 silos for its silo-based ICBMs.[5] This development will increase China's already existing 20 silos that are operational for its liquid fuelled DF-5 ICBMs. Reportedly, these silos are being constructed at three silo fields near Yumen in Gansu province in north-central China, another one near Hami city in eastern Xinjiang and the latest one is likely in Inner Mongolia, with a possible 29 silo construction sites. This huge silo construction is likely to further escalate military tension in the region. It is predicted that upon operationalisation, the number of China's ICBM launchers might exceed the US' presently deployed 400 launchers of Minuteman III ICBMs.[6]

In the last decade, China has added two new types and two new variants of land-based ballistic missiles in its inventory. This addition includes the DF-5B, DF-5C, DF-26, DF-41. While some of these missiles were developed and tested prior to 2010, they were either deployed or publicly introduced after 2010. Based on available data, the ICBM DF-41 has been tested ten times since 2012. The DF-26 has been tested twice since January 2019; the Submarine Launched Ballistic Missile (SLBM) JL-2 has also been tested twice in the last ten years.

Pakistan is no exception; it continues to expand its nuclear arsenal with more warheads and delivery systems. According to an estimate by the *The Bulletin of the Atomic Scientists*, as of 2021, Pakistan had a nuclear weapon stockpile of 165 warheads, which is estimated to grow to 200 warheads in the next five years.[7] Whereas, in 2010, the country had an

5. Matt Korda and Hans Kristensen, "China Is Building A Second Nuclear Missile Silo Field", Federation of American Scientists, July 26, 2021. https://fas.org/blogs/security/2021/07/china-is-building-a-second-nuclear-missile-silo-field/. Accessed on October 16, 2021.
6. Rod Lee, "PLA Likely Begins Construction of an Intercontinental Ballistic Missile Silo Site near Hanggin Banner", China Aerospace Studies Institute. https://www.airuniversity.af.edu/CASI/Display/Article/2729781/pla-likely-begins-construction-of-an-intercontinental-ballistic-missile-silo-si/. Accessed on August 12, 2021.
7. Hans M. Kristensen and Matt Korda, "Nuclear Notebook: How Many Nuclear Weapons Does Pakistan Have in 2021?", *The Bulletin of the Atomic Scientists*. https://thebulletin.org/premium/2021-09/nuclear-notebook-how-many-nuclear-weapons-does-pakistan-have-in-2021/. Accessed on September 7, 2021.

estimated nuclear weapons stockpile of 90-110 nuclear warheads. With this growth in nuclear warheads, several delivery systems have also been developed.

Presently, the country has six types of nuclear capable ballistic missiles: the Shaheen series of missiles, Ababeel, Ghauri, Ghaznavi, Abdali, and Nasr. Three of these missile variants are under development, i.e., Shaheen III, Shaheen 1A and Ababeel. Based on the launch platform, it has three types of cruise missiles: the Babur series of Ground-Launched Cruise Missiles (GLCMs), Babur and Babur-2; Submarine-Launched Cruise Missiles (SLCMs), Babur-3; and Air-Launched Cruise Missiles (ALCMs), Ra'ad and Ra'ad-II; of these, the SLCM Babur-3, GLCM Babur-2/1 (B) and ALCM Ra'ad-2 are undergoing development. Apart from these, it is speculated that the ICBM Taimur, with a range of 7,000 km, is under development by Pakistan.

In the last decade, Pakistan has added seven kinds of missiles to its missile inventory. These additions include the Nasr missile in 2011, Shaheen-1A in 2012, Shaheen-3 in 2015, Babur-2 cruise missile in 2016, Ababeel, Babur-3 missiles in 2017 and Ra'ad-2 missile in 2020. Missiles which were tested frequently are the Nasr (eight times), Ghaznavi (six times), Abdali (three times), Shaheen 1A (three times), Ghauri (four times), Babur and Babur-2 (six times) and Ra'ad and Ra'ad 2 (four times).

India's nuclear missile capability has been growing to address its security concerns emerging from both its nuclear neighbours. Presently, India has nine types of nuclear capable ballistic missiles with their land and sea-based variants, i.e., the Prithvi-II, Agni-I, Agni-II, Agni-III, Agni-IV, Agni-V, Dhanush, K-15 and K-4. Among these missiles, the Dhanush, K-4 and K-15 are sea-based ballistic missiles. The Dhanush and Prithvi-II are all liquid fuelled missiles while the other missiles are solid propellant systems. Also, the Agni-IV and Agni-V ballistic missiles are presently under development by the Defence Research and Development Organisation (DRDO). Reportedly, another longer-range land-based ballistic missile, the Agni-VI, that has a "force multiplier capability by the MIRV approach", is under development by DRDO.[8] Similarly, it is reported that a nuclear capable hypersonic missile the Shaurya, is under development by DRDO and has been flight tested since 2011[9], although there are no government sources available regarding the development.

In the last decade, India has added one land-based ballistic missile, e.g., the Agni-III and two sea-based nuclear capable ballistic missiles, e.g., the Dhanush and K-15, to its nuclear missile inventory. Besides this, the nuclear-powered ballistic missile submarine (SSBN) INS *Arihant* was also made operational in November 2018. Another nuclear- powered ballistic missile submarine, the INS *Arighat* was launched in 2017 but is yet to be fully operationalised.

Missiles that were tested frequently in this time period are the Prithvi-II (26 times), Agni-I (11 times), Agni-II (seven times), Agni-III (seven times), Agni-IV (seven times), Agni-V (seven times), K-15 (six times) and Dhanush (five times).

Reportedly, India's nuclear warheads have increased from 60-80 warheads in 2010 to 150 warheads in 2020.[10] Both its nuclear neighbours, China and Pakistan continue to have a larger number of nuclear warheads than India.

---

8. PTI, "India Developing Agni-VI Missile Capable of Carrying Multiple Warheads: DRDO Chief", *Business Line*, March 9, 2018. https://www.thehindubusinessline.com/news/india-developing-agni-vi-missile-capable-of-carrying-multiple-warheads-drdo-chief/article20575500.ece1.
9. Hemant Kumar Rout, "India Successfully Test-Fires New Nuclear Capable Hypersonic Missile Shaurya", *The New Indian Express*, October 3, 2020. https://www.newindianexpress.com/nation/2020/oct/03/india-successfully-test-fires-nuclear-capable-hypersonic-missile-shaurya-2205284.html; also, T.S. Subramanian and Y. Mallikarjun, "India Successfully Test-fires Shourya Missile", *The Hindu*, September 24, 2011. https://www.thehindu.com/sci-tech/science/india-successfully-testfires-shourya-missile/article2482010.ece
10. Hans M. Kristensen and Matt Korda, "Indian Nuclear Forces, 2020", *The Bulletin of the Atomic Scientists*, Vol 76, Issue 4, 2020.

## Dual Capable Missiles

China has many missiles, including cruise missiles, that are speculated to be dual capable. China's land-based ballistic missile, the DF-15, has been declared as nuclear capable many times in the past, though the 2020 US DoD report lists it as a conventional missile. Some of its cruise missiles are also likely to have nuclear capability, but China has always maintained ambiguity on this matter in order to add to its deterrent value. The 2021 US DoD report confirms that the solid fuelled DF-26 is designed to rapidly swap conventional and nuclear warheads and is capable of conducting precision strikes against ground targets, as well as conventional strikes against naval targets. While this ambiguity is an inherent part of China's nuclear policy, it means complex challenges for other nuclear weapon countries as the presence of uncertainty related to the nature of the payload might lead to miscalculations and nuclear escalation.

Pakistan is also reported to have dual capable missiles. Its air-launched cruise missile, the Ra'ad, sea-based cruise missile, the Babur (Hatf-7) and land-based ballistic missile, the Shaheen-I (Hatf-4) are dual-capable missiles.[11] Apart from these, Pakistan claims its short-range ballistic missile, the Abdali (Hatf-II), is dual capable and claims that it "carries nuclear as well as conventional warheads with high accuracy"[12]. All of these missiles are in the relatively short-range category, i.e., the Ra'ad has a range of 350 km, the Babur has a stated range of 600 km to 700km, but US intelligence sources state the range as 350 km, the Abdali has a range of 180 km and the Shaheen-I has a maximum range of 650 km. The short-range missiles provide Pakistan with the advantage of targeting and deployment options in the conventional and nuclear modes to primarily target India to limit its conventional force superiority vis-a-vis Pakistan in the event of a crisis. Clearly, these actions have the

11. Hans M. Kristensen and Matt Korda, "Pakistani Nuclear Weapons, 2021", *Nuclear Notebook, The Bulletin of the Atomic Scientists,* Vol. 77, Issue 5, 2021, DOI: 10.1080/00963402.2021.1964258.
12. ISPR Report No PR-20/2013-ISPR, February 15, 2013. https://www.ispr.gov.pk/press-release-detail.php?id=2242.

potential to lower the threshold of the conflict and lead to dangerous escalation owing to uncertainty related to the nature of the warheads.

India maintains clarity in defining its nuclear and conventional missiles. All its cruise missiles are conventional in nature. There is speculation that India's Nirbhay cruise missile is dual capable, however, there is no official confirmation from the government or the US intelligence community.[13] Therefore, it does not create ambiguity related to the nature of the warheads that is caused by dual capable missiles.

## MIRV and MaRV Capability

China perceives the US missile defence architecture as a major challenge to its own missile operations. As a result, China has resorted to improving the penetrability of its missiles against adversary missile defences by investing in development measures such as MIRVs and MaRVs.

Presently, China's DF-5B ICBMs have MIRV capability. A MIRVed missile can carry several miniaturised warheads, each of which can be targeted independently, thereby providing a greater advantage over a missile containing a single warhead. The DF-5B ICBM is liquid-fuelled and silo-based with a range of 13, 000 km that can reach the continental US. While the long time required for fuelling the missile makes it more susceptible to an adversary attack, the missile's intercontinental range, greater throwweight and most significantly, the capability to carry a MIRVed warhead increases its ability to penetrate missile defences and enhances the capability of China's missile force with more survivable delivery systems. In addition to this, a new generation of silo-based/road mobile, solid-fuelled ICBM, the DF-41 that is undergoing development in China, reportedly has a greater number of MIRVed warhead carrying capabilities.

Another ICBM, an upgraded version of the DF-5B, ICBM DF-5C, with the stated capability of carrying 10 MIRVs, is in development and was flight-tested in February 2017. These developments demonstrate China's increased emphasis on developing MIRVed missiles with

13. According to Kristensen and Korda, n. 10.

improved warheads, as well as a strong preference for developing more solid-fuelled missiles while maintaining a limited number of liquid-fuelled missiles to increase the credibility of its deterrence.

According to the 2021 US DoD report, China's conventionally armed DF-21D is fitted with a MaRV warhead, and is capable of rapidly reloading in the field. MaRVed warheads are capable of performing pre-programmed flight manoeuvres during the atmospheric reentry phase. They are also difficult to intercept and, therefore, better able to engage adversary missile defence systems.[14]

Pakistan's solid-fuelled Ababeel missile is claimed to have MIRV capabilities. Pakistan's claim of MIRV technology has not been verified and remains a point of debate. MIRV technology requires a broad degree of technical sophistication and there is strong evidence that Pakistan might have taken assistance from China to develop the technology.[15] Pakistan cites India's Ballistic Missile Defence (BMD) development as the rationale for its MIRV capability development. However, India's BMD systems are in the early stages of development and are intended to provide limited coverage rather than nationwide coverage, hence, this development of the defensive system is unlikely to change the strategic equation with Pakistan. The MIRV capability provides the attacking state with a counter-force strike capability. Land-based MIRV systems indicate essentially a first strike capability, and during a crisis, deployed MIRVed missiles could have a negative impact on deterrence stability by increasing the temptation for preemption.

Reportedly, India's DRDO is developing the Agni-VI ballistic missile that has a "force multiplier capability by the MIRV approach"[16]. Although there is no official confirmation on this, India's present development

14. US-China Economic and Security Review Commission, *China's Offensive Missile Forces*, Chapter 2, Section 3, Annual Report, 2015. https://www.uscc.gov/annual-report/2015-annual-report-congress. Accessed on October 15, 2020.
15. Stephen Chen, "China Provides Tracking System for Pakistan's Missile Programme", *The South China Morning Post*, March 22, 2018. https://www.scmp.com/news/china/society/article/2137643/china-provides-tracking-system-pakistans-missile-programme.
16. n. 8.

seems to be focussed on developing the Agni-V missile to address its threat perception from China, and developing and deploying Multiple Reentry Vehicles (MRVs) and MaRVs to potentially build capabilities against China's BMD systems.[17]

### Hypersonic Weapon Programme

China's hypersonic weapon programme is rapidly advancing and it has operationalised the DF-17 solid-fuelled missile equipped with the Hypersonic Glide Vehicle (HGV) DF-ZF that can reach speeds of Mach 5 to Mach 10 (approximately 6,174 to 12,348 km per hour) in its glide phase, which is far greater than the speed that missile defences are built to counter.[18] China reportedly conducted nine tests of the HGV between January 2014 and November 2017 and the missile has demonstrated a high degree of accuracy.[19]

While hypersonic weapons are effective against ballistic missile defence systems, China is yet to fully develop this technology and has intensified research on developing and testing HGVs in recent times. Reportedly, in August 2021, China tested its first ever "nuclear capable hypersonic missile"; the missile circled the globe before missing its target by a few miles, "demonstrating an advanced space capability that caught US intelligence by surprise"[20]. China has denied the missile test and described it as a hypersonic "vehicle" test. As China is not bound by any arms control deals and has been unwilling to do so, this new development,

---

17. Manpreet Sethi, "Missile Developments in Southern Asia: A Perspective from India", *IISS*, June 17, 2021. https://www.iiss.org/blogs/research-paper/2021/06/missile-developments-southern-asia.
18. Missile Defence Project, "DF-17, Missile Threat," Centre for Strategic and International Studies, February 19, 2020. Last modified June 23, 2020, https://missilethreat.csis.org/missile/df-17/. Accessed on October 15, 2020.
19. Ibid.
20. PTI, "India Among Select Few Countries Developing Hypersonic Missiles: US Congressional Report", *The Hindu*, October 22, 2021. https://www.thehindu.com/news/international/india-among-select-few-countries-developing-hypersonic-missiles-us-congressional-report/article37119764.ece.

along with China's growing dominance in other missile technologies, may have a long-term impact on regional stability.

Pakistan has not demonstrated hypersonic capability yet and is not known to have an indigenous hypersonic technology programme. It is possible that Pakistan will acquire this technology in the future with China's assistance, as it did in the case of MIRV technology and several other nuclear missiles. Similarly, some Pakistani analysts argue that part of Pakistan's "emerging menu of long-range [anti-access, area denial] capabilities are increasingly going to be needed for maintaining a credible deterrent"[21].

Although China claims its nuclear missile programme is primarily US-centric, the active modernisation and deployment of MIRV and BMD technologies have direct implications for India. In view of the threat perception from these developments, India is also engaged in Research and Development (R&D) of hypersonic technology. On September 7, 2020, the DRDO test-fired a Hypersonic Technology Demonstrator Vehicle (HSTDV) using a solid rocket motor that will lead to the development of a hypersonic cruise missile system in the future.[22] An earlier test of the vehicle was conducted in June 2019. The DRDO chief mentioned that "it will take around four to five years... to realise the complete missile system with some good amount of range" and a speed of around 6 to 7 Mach.[23] Another component of India's hypersonic capability is the Hypersonic Cruise Missile (HCM). In collaboration with Russia, India is developing the HCM BrahMos II with a speed of 7 Mach. According to the 2021 US Congressional Research Service (CRS)

21. Mike Yeo, Nigel Pittaway, Usman Ansari, Vivek Raghuvanshi and Chris Martin, "Hypersonic and Directed-Energy Weapons: Who has Them, and Who's Winning the Race in the Asia-Pacific?", *Defence News,* March 15, 2021. https://www.defensenews.com/global/asia-pacific/2021/03/15/hypersonic-and-directed-energy-weapons-who-has-them-and-whos-winning-the-race-in-the-asia-pacific/.
22. ANI, "India Can Have Complete Hypersonic Cruise System in 4-5 Years: DRDO", *The Economic Times,* October 14, 2020. https://economictimes.indiatimes.com/news/defence/india-can-have-complete-hypersonic-cruise-missile-system-in-4-5-years
23. Ibid.

report on hypersonic weapons, the missile is expected to achieve its initial operational capability between 2025 and 2028.[24]

Presently, India's hypersonic development programme remains in the R&D stage. This development of hypersonic capability by India comes at a time when only the US, Russia and China have the capability and are developing hypersonic glide vehicles. It seems India finds it prudent to develop the technology and gain leverage in the region at a time when a number of other countries, including Japan, are developing the capability.[25] The development is a major breakthrough in India's missile development programme. The scramjet engine would enable the cruise missile to travel at hypersonic speed while withstanding high temperatures. The test successfully demonstrated many critical technologies such as "aerodynamic configuration for hypersonic manoeuvres, use of scramjet propulsion for ignition and sustained combustion at hypersonic flow, thermo-structural characterisation of high temperature materials and separation mechanisms at hypersonic velocities"[26]. It is likely that India will further develop the technology.

## Sea-Based Platforms

Due to its stealth and endurance capability, the survivability of a nuclear-powered submarine as a platform for SLBMs and SLCMs makes it the most reliable component of the nuclear triad and offers an assured second-strike capability.

China's solid-fuelled Julang/JL-2 is an SLBM with a range of 7,000 km. Although the missile was first launched in 2002, it suffered a series

24. Kelley M. Sayler, "Hypersonic Weapons: Background and Issues for Congress", Congressional Research Service, October 19, 2021. https://sgp.fas.org/crs/weapons/R45811.pdf.
25. n. 20.
26. Ministry of Defence, "DRDO Successfully Flight Tests Hypersonic Technology Demonstrator Vehicle", Press Information Bureau (PIB), September 7, 2020. https://pib.gov.in/PressReleasePage.aspx?PRID=1651956.

of test failures until its first successful test in August 2012.[27] The missile is deployed on China's four operational type 094 Jin class SSBNs which are equipped to carry up to twelve JL-2 SLBMs. According to the US DoD report 2021, a longer range upgraded variant of the SLBM-JL-3 is presently under development and it will likely be deployed on China's next-generation 096 SSBN. The 096 SSBN is presently under development and it is expected to be quieter than the current Jin class SSBN.[28] Some analysts predict that the new JL-3 is estimated to have a range of more than 9,000 km, with MIRV capability.[29] The first test of the SLBM was conducted in November 2018. With the development of the JL-2, China acquired its first credible sea-based nuclear deterrent that would enable it to have a viable sea-based second-strike capability. It will also help it to build a nuclear triad of delivery systems that would be completed with the development of a nuclear capable Air-Launched Ballistic Missile (ALBM) that is reported to be undergoing development by China.

In Pakistan, the submarine-based nuclear capability is managed by the Headquarters Naval Strategic Forces Command (NSFC), which the government said in 2012 would be the "custodian of the nation's second strike capability"[30]. Pakistan is developing a Submarine Launched Cruise Missile (SLCM), the Babur-3, with a range of 450 km. The missile is under development and the first test of the missile

---

27. Hans M. Kristensen and Robert S. Norris, "Chinese Nuclear Forces, 2011", *The Bulletin of the Atomic Scientists*, Vol 67, Issue 6, pp. 81-87. Also see, Annual Report to Congress: Military and Security Developments Involving the People's Republic of China 2013. www.defence.gov/pubs/2013_china_report_final.pdf.
28. Rajaram Nagappa and N Ramamoorthy, "Changing Nature of Deterrence: The Challenge of Asymmetric Threats", *National Security*, Vivekananda International Foundation, Vol. III, Issue-2, 2020, pp. 213-231.
29. Matthew P. Funaiole, Joseph S. Bermudez Jr. and Brian Hart, "A Glimpse of Chinese Ballistic Missile Submarines", Centre for Strategic and International Studies (CSIS), August 4, 2021. https://www.csis.org/analysis/glimpse-chinese-ballistic-missile-submarines.
30. Hans M. Kristensen, Robert S. Norris and Julia Diamond, "Pakistani Nuclear Forces", *The Bulletin of the Atomic Scientists*, Vol 74, Issue 5, 2018, pp. 348-358, DOI: 10.1080/00963402.2018.1507796

was conducted in January 2017; another test followed in March 2018.[31] Some studies have claimed that the missile was most likely launched from the diesel-electric Agosta-90B (Khalid class) submarine in service with the Pakistan Navy.[32]

Given the fact that Pakistan does not have a nuclear-powered submarine, the potential deployment of the SLCM and its ability to work at full capacity might be affected as the operational superiority of a nuclear-powered submarine over a diesel-powered submarine makes it a more effective launch platform for submarine-launched missiles. However, in 2016, Pakistan signed a deal with China to buy eight Yuan class diesel-electric attack submarines. It was also reported that these submarines will have the air independent propulsion system.[33] The submarines are expected to be completed between 2023 and 2028. These submarines will be a significant addition to Pakistan's submarine fleet, considering the fact that many modern attack submarines can launch cruise missiles while remaining underwater and their air independent propulsion systems could significantly contribute to effective stealth capability. Apart from this, Pakistan may also be seeking to build a nuclear-powered submarine, although it is unclear if any work on this front has taken place yet.[34]

As sea-based platforms for its SLBMs—the K-4 and K-15 (Sagarika)—and ship-based ballistic missile, Dhanush, India operates the SSBN, INS *Arihant*, and specially configured Sukanya class vessels, the *Subhadra* and *Suvarna*. Another SSBN, the INS *Arighat* is undergoing development and it is speculated that two more SSBNs, the

31. ISPR Report No PR-10/2017-ISPR, January 9, 2017. https://www.ispr.gov.pk/press-release-detail.php?id=3672
32. Rajaram Nagappa et al., "Babur-3—Pakistan's SLCM: Capability and Limitations", *Air Power Journal,* Vol. 13 No. 3, Monsoon 2018 (July-September). Accessed on January 5, 2020.
33. Zia Mian, M.V. Ramana and A.H. Nayyar, "Nuclear Submarines in South Asia: New Risks and Dangers", *Journal for Peace and Nuclear Disarmament*, 2019, 2:1, pp. 184-202, DOI: 10.1080/25751654.2019.1621425
34. Ibid.

S-4 and S-4*, will enter service by 2024.[35] It is also predicted that a new S-5 class submarine might be under development by India, but there is no official confirmation on this matter.

Among India's SLBMs, the K-4 was tested for the first time in March 2014 and the missile is awaiting induction into the force. The INS *Arihant* was commissioned in August 2016 and it successfully completed the first deterrence patrol on November 6, 2018, before becoming operational. Another SSBN, the INS *Arighat* was launched in November 2017. Clearly, significant modernisation and development have taken place in the last decade in India's SSBNs and SLBMs. It is predicted that in the coming years, India will further strengthen its sea-based deterrence by adding the improved new SSBNs and SLBMs to its fleet.

India, China and Pakistan aspire to strengthen their sea-based deterrent with a common aim of achieving credible second-strike capability. India and China, with a No First Use (NFU) policy, need the capability even more. None of these countries has yet achieved the complete nuclear triad that is perceived to be essential for full spectrum deterrence. It is likely that they will further develop the capability.

These trends indicate that all these countries are investing in new and sophisticated nuclear technologies and capability building to reduce threat perceptions from the steady and continuous modernisation of neighbouring countries. Dialogue on strategic issues and effective arms control measures can be a few options to arrest some of the destabilising developments in the region. Also, the US could play a significant role in this as developments in US-China strategic relations directly influence the dynamics of Southern Asia.

35. Kristensen and Korda, n. 10.

# 11. The Threat of N-Terrorism: A Contemporary Assessment

*Sitakanta Mishra*

## The Changing Context

The world has no precedent yet of a successful nuclear terror event but two decades after 9/11, preventing and countering terrorism remains one of the most challenging tasks for many nations. Though the likelihood of full-fledged nuclear terrorism is often dismissed, "there are examples of both threats and feasible pathways to conduct nuclear terrorism."[1] It will remain a critical security concern on the global agenda for the foreseeable future, largely due to the singularly destructive power of nuclear weapons, and the evolving geopolitical dimension of the threat. This has been further exacerbated after the upsurge of the Taliban and Al-Qaeda, and their nexus with other terrorist groups nurtured by nuclear weapon states that have also adhered to "terror as an instrument of foreign policy."[2]

The role nuclear weapons played in the realm of global security entails that the world has experienced 75 years of nuclear peace, without the bomb being used again. But the underlying assumption that nuclear weapons are the ultimate source of national security needs careful reevaluation owing to the dangers inherent in their possession,

1. Elisabeth Suh and Katharina Miriam Hess, "Nuclear Terrorism: A Plausible and Pestilent Threat", *DGAP Online Commentary*. https://dgap.org/en/research/publications/nuclear-terrorism, September 8, 2021.
2. Grant Wardlaw, "Terror as an Instrument of Foreign Policy", *Journal of Strategic Studies*, 10: 4, 1987, pp. 237-259.

management and safe-keeping. Strategists have long feared the possibility of leakage of nuclear weapons to terrorists. Given the rapid proliferation and attractiveness of nuclear technology, it is assumed that terrorists may use the terror potential of nuclear weapons for their nefarious designs. The 9/11 attack has already demonstrated the capability and reach of terrorists. Today, extremist organisations (or non-state actors) with a terrorism connection and non-democratic ideology have come to power in states; nuclear weapons may not be far away from their possession if they desire to acquire them. Therefore, the nuclear discourse seems to be moving from the era of *security by nuclear weapons* to *security of nuclear weapons.*

This chapter examines the threat to nuclear weapons and the probability of nuclear terrorism by analysing different options available for the terrorists to exploit the existing loopholes in the global nuclear governance. It explores the changing global strategic situation and the possibility of nuclear terrorism. The chapter delves into the evolving threat scenario with the upsurge of the Taliban and Al-Qaeda, and the nexus between terrorist networks and nuclear weapon states. Keeping in mind the fact that in Afghanistan, non-state actors are now at the helm of affairs, or custodians of the state, with the backing of another nuclear weapon state, Pakistan, this study prescribes the creation of a global consensus on the urgency of the threat of nuclear terrorism and actions to mitigate the menace.

## From 'Security By' to 'Security Of' Nuclear Weapons

The conventional wisdom on nuclear weapons as purely national security assets necessitates a second thought, especially in the post-9/11 era. Undoubtedly, the end of the Cold War has dramatically reduced the risk of nuclear war, but the legacy of the nuclear danger continues in many other forms, posing serious challenges to international security. The major challenge is the security of the nuclear weapons and materials. Thomas Kean, the chairman of the 9/11 Commission, said, "A nuclear weapon in the hands of a terrorist is the single greatest threat...today."[3] The former US Energy Secretary Bill Richardson also voiced his apprehension on the misuse of nuclear technology, saying that "in the 20th century, nuclear deterrence worked. In the 21st century, it won't. MAD (Mutually Assured Destruction) deterred the Soviet Union, but nothing will stop suicidal *jihadists* from using a nuclear bomb if they get their hands on one."[4] Also it is quite clear by now that most of the threats of the 21st century emanate not from states but from societies—from social trends and fault lines. Does today's nuclear discourse adequately address the societal linkage of nuclear threats?

In recent times, the state patronage of terrorists and the politico-strategic conditions in nuclear weapons possessing countries "have converted nuclear weapons from queens on the chessboard of international politics into military and political hostages."[5] Nuclear weapons are acquired by states in the name of strengthening national security. But proclaiming one's assets are absolutely secured, says Pervez Hoodbhoy, is simply "an act of supreme folly by creating a false sense security".[6] The apprehension of a nuclear terror incident has become more pronounced in the post-9/11 period. In August 2001, the Pakistani ISI (Inter-Services Intelligence) had proof that Osama bin Laden had received in person two retired nuclear scientists, Sultan Bashiruddin Mahmood and Chaudhiri Abdul Majeed, at his secret headquarters in Afghanistan.[7] Also documents found in Afghanistan reveal that Al-

---

3. Quoted in Charles B. Curtis, "Promoting Global Best Practices," Institute for Nuclear Materials Management, 46th Annual Meeting, July 11, 2005, p. 2. URL: http://www.nti.org
4. Bill Richardson, "Preventing a Nuclear 9-11." URL: http://www.richardsonforpresident.com, March 28, 2007. Accessed on March 29, 2007.
5. Stephen J. Cimbala, "Nuclear Weapons in the New World Order," *The Journal of Strategic Studies*, Vol. 16, No. 2, June 1993, p. 184.
6. Pervez Hoodbhoy, "Is Accidental Nuclear War Impossible?", December 8, 1998. http://www.chowk.com/articles/4419
7. Adrian Levy and Catherine Scott-Clark, *Deception: Pakistan, The United States and The Global Nuclear Weapons Conspiracy* (New Delhi: Penguin Books India, 2007), pp. 309-310.

Qaeda's efforts to make nuclear weapons were far less sophisticated, but the group is determined to get them and willing to use them.[8] One may argue that both Al-Qaeda and other major *jihadist* terror groups have suffered tremendous blows because of the US-led counter-terrorism operations, therefore, the possibility of a terrorist nuclear attack need not be worried about. But the US withdrawal from Afghanistan and consequent regional uncertainty involving the "Taliban-Pakistan nexus paint a more worrisome dimension of the possibility of the Taliban acquiring nuclear weapons from Pakistan—a nightmarish scenario."[9] In a letter dated August 24, 2021, signed by the group of 68 US lawmakers from the Senate and the House of Representatives, they have asked the Biden Administration: "What is your plan to help to ensure that the Taliban does not destabilize its nuclear neighbour, Pakistan? Do you have a plan to ensure that Afghanistan, under Taliban occupation, will never acquire a nuclear weapon"?[10]

Furthermore, the US senators have introduced a legislation titled "Afghanistan Counterterrorism, Oversight, and Accountability Act" seeking sanctions on the Afghan Taliban and the foreign governments that support the outfit, referring to Pakistan's role. Even US Army General Mark A. Milley has reportedly told the Senate Armed Services Committee that President Biden was warned that accelerated withdrawal from Afghanistan would increase the risks of regional instability and could jeopardise Pakistan's nuclear arsenal.[11] In a report published in the

8. David Albright, *Al Qaida's Nuclear Programme: Through the Window of Seized Documents*, Nautilus Institute Special Forum 47, November 6, 2002. URL: www.nautilus.org/fora/special-Policy-Forum/47_Albright.html#sect2
9. SD Pradhan, "Pak Nukes with Taliban: Deserves Serious Attention of the World Community", *The Times of India*, October 1, 2021.
10. "US Lawmakers Write to Joe Biden Demanding Answers Over Afghanistan, China, Terrorism, Nuclear Weapon and More". https://www.wionews.com/world/us-lawmakers-write-to-joe-biden-demanding-answers-over-afghanistan-china-terrorism-nuclear-weapon-and-more-409006, August 28, 2021.
11. Statement of General Mark A. Milley, USA, 20th Chairman of the Joint Chiefs of Staff Department of Defence Afghanistan Hearing, September 28, 2021. https://www.armed-services.senate.gov/imo/media/doc/Printed%2028%20Sep%20SASC%20CJCS%20Written%20Statement.pdf

*Washington Post*, John R. Bolton, who served as the national security adviser under Donald Trump, argued that "the prospect that Pakistan could slip individual warheads to terrorist groups to detonate anywhere in the world would make a new 9/11 incomparably more deadly."[12] Bolton stated that the "Taliban could obtain nuclear weapons from Pakistan", pointing out that "parts of the Pakistani government are directly responsible for the Taliban's return to power." Moreover, the possibility of the Islamic State of Iraq and Syria-Khorasan (ISIS-K) acquiring nuclear weapons from Pakistan cannot be ruled out. In the worst case, "what if internal Taliban take over nuclear assets?" All these indicate that the renewed threat of nuclear terrorism is unfolding in a new *avatar*, therefore, must be dealt with preemptively and multilaterally.

## Why Terrorists would Seek Nuclear Weapons?

Since its inception, nuclear technology has evoked a great sense of fear (along with a sense of awe) because of its immense destructive potential. The fear and allure of the atom bomb's power instigated nations to acquire them. The same factor is equally attractive for terrorists. There is a phenomenal rise in the number and capability of terrorist groups. Though many terrorist groups in the past have threatened to resort to the nuclear path, probably they could not overcome their conservativeness regarding the tactics and level of violence in their pursuit. But the newer terrorist groups in the post-Cold War period seem to be more violent, more innovative, and more willing to use mass-killing weapons.[13] More importantly, the nexus between terrorism and politics has become entrenched with the patronage of terrorists and adoption of terrorism as a state-policy by some states; the recent upsurge of the Taliban, backed by a nuclear-armed Pakistan, which is now at the helm of affairs in Afghanistan is just one dimension of the threat.

---

12. John R. Bolton, "The Time for Equivocating About a Nuclear-Armed, Taliban-Friendly Pakistan is Over". https://www.washingtonpost.com/opinions/2021/08/23/john-bolton-taliban-takeover-pakistan-extremists/, August 23, 2021.
13. "Terrorism: Motivations and Causes". http://www.terrorism.com/modules.php?op=modload&name=News&file=article&sid=5693&mode=thread

The specific answers to the question of 'why' terrorists might resort to nuclear weapons are equally logical: (1) While a state possessing a nuclear weapon can effectively deter its adversary, it cannot deter a non-state actor since it doesn't have a return address. But a terrorist organisation possessing a nuclear arsenal can bring a state to its knees. With the possession of the ultimate weapon, a terrorist outfit can effectively blackmail a state and bargain for its cause from the ultimate platform of strength. "Apocalyptic visions or global ambitions drove groups such as Al-Qaeda and the Japanese terror cult Aum Shinrikyo to seek nuclear weapons."[14] (2) Nuclear terrorism has the feature of achieving a unique type of public fear and trauma because of the negative public perception on anything nuclear. The public's irrational fear of radiation can well be exploited by terrorists. (3) The prestige and unique identity factor involved in the possession of nuclear weapons would instigate a terrorist organisation to acquire them. By acquiring nuclear capability, a terrorist outfit could assume a quasi-state nature. That would distinguish the entity from other entities since the mere identity of 'being nuclear' would certainly ensure that it had vast coercive power, with the ability to have considerable impact on the public and government. (4) The popular fear of nuclear weapons and the panic involved with such an act appropriately matches the aims and objectives of terrorists. The nuclear weapon today is highly feared and the mere news of its possession by any non-state actor would create public panic. (5) Lastly, the global nuclear discourse that revolves around state behaviour, needs a fundamental rethinking now when non-state actors have emerged as rivals of state actors and have the "motives, means and opportunities" to secure, prepare and use a nuclear bomb.[15]

---

14. Matthew Bunn et al., "Revitalizing Nuclear Security in an Era of Uncertainty", https://www.belfercenter.org/publication/revitalizing-nuclear-security-era-uncertainty, January 2019.
15. Ibid., pp. 3-4.

## No Precedent but Attempts

To the question, is there any precedent of a successful nuclear terror incident, the answer is simply 'No'. But to the question, have terrorists pursued nuclear weapons, the answer is definitely 'Yes'. The terrorist cult Aum Shinrikyo released the Sarin nerve gas in Matsumoto and in the Tokyo subway in 1995 and attempted to acquire both nuclear and biological weapons. Al-Qaeda, whose leader declared acquisition of nuclear and chemical weapons to be a "religious duty," had a focussed nuclear weapons effort that reported directly to Ayman al-Zawahiri. Al-Qaida, as revealed by documents seized in Afghanistan, actively sought nuclear weapons and clearly expressed the desire to use them if it was able to acquire them. This effort included repeated attempts to get nuclear material and recruit nuclear expertise, and progressed as far as carrying out crude but sensible tests of conventional explosives in the Afghan desert. Chechen terrorists planted a stolen radiological source in a Moscow park as a warning, repeatedly threatened to sabotage nuclear reactors, and reportedly carried out reconnaissance on both nuclear weapon storage sites and nuclear weapon transport trains. So far, there is no public evidence of a focussed Islamic State (IS) effort to acquire nuclear weapons, despite some hints, including video monitoring of the home of a top official of a Belgian nuclear research centre.

Moreover, there have been many documented cases of misappropriation of plutonium or Highly Enriched Uranian (HEU) during 1992-2019, though these incidents involved quantities not large enough to make a nuclear weapon. But they constitute empirical confirmation of nuclear security failures, resulting in loss of control of fissile material. In a Belfer Centre Report (January 2019), Mathew Bunn and others have listed a number of such attempts in recent times in various parts of the world:

- On November 8, 2007, four armed men broke into South Africa's Pelindaba nuclear facility. At the time, Pelindaba housed hundreds of kilograms of weapon-grade uranium, enough to construct an estimated 25 nuclear bombs.

- Russia: General-Major Victor Gaidukov, commander of a nuclear weapon storage site, was fired over accusations of accepting $300,000 in bribes (2010).
- Pakistan: Brigadier General Ali Khan was arrested for ties to Islamic extremists (2011).
- Belgium: Peace activists broke into nuclear weapon storage base, and spent over one hour there before being detected and stopped (2010).
- In 2012, explosives were found under a truck at the Ringhals nuclear power plant, the largest in Sweden. Fortunately, the explosives were not connected to a detonator.
- In 2013, two people scaled the fence at Belgium's HEU research reactor, broke into the facility, and stole equipment.
- In 2014, a computer in the control room (though not one actually controlling the reactor) at Japan's Monju nuclear reactors was hacked.
- In 2016, the Belgian nuclear agency's computer system was hacked and forced to briefly shut down.
- In 2017, Greenpeace activists twice penetrated security barriers at French nuclear power plants to protest against nuclear energy and highlight what they asserted were security weaknesses.

The report also identified a number of worrisome incidents involving US tactical nuclear weapons stationed in Europe which suggested, that "tactical nuclear weapons systems present a particular risk of theft, especially when they are out of garrison."[16] The world should be concerned by the increased security challenges that accompany growing stockpiles, particularly of Tactical Nuclear Weapons (TNWs) that are designed for use on the battlefield. And these systems are a source of concern because they're susceptible to theft due to their size and mode of employment. Amidst volatility and upsurge in extremism in the neighbourhood, Pakistan is increasing its TNW stockpile.

During the Cold War, the level of alert was high on several occasions, often owing to human or technical errors. More than three decades after

16. Ibid., pp. 30-31.

the end of the Cold War, 3,500 nuclear warheads still remain on hair-trigger alert.[17] Additionally, nuclear technology has spread to countries with unstable regimes in conflict zones. Though there is no recent news of nuclear misuse or misappropriation, the threat of a nuclear nightmare remains strong as the entire process—from fabrication till their deployment/stockpile and command and control—involves enormous risks owing to the fallibility of human and mechanical factors. Seventy-five years later and 30 years after the end of the Cold War, the threat of a nuclear nightmare remains strong.

### Feasibility and Pathways

Through certain pathways, the non-state actors can have access to nuclear technology and weapons. Many believe that terrorists may be able to prepare a radiation dispersal device or "dirty bomb" but fabrication of a full-fledged nuclear device by them would not be feasible as it requires state-level resources, knowhow and a huge establishment. At best, they are capable of sabotaging a nuclear facility, or a 9/11-type of attack, to cause damage or take possession of warheads. There have been such attempts in the past. For example, the Al-Qaeda's senior leadership had explored the possibility of sabotaging nuclear facilities; Chechen terrorists had threatened and planned attacks on nuclear facilities. But modern-day nuclear arsenals are highly secured assets, guarded by layers of security and managed by well-planned command and control systems; in peace-time, generally, they are kept disassembled and designed with utmost care to explode only when deliberately armed and fired.

Until now it has been believed that terrorists would not be able to get hold of a ready-to-use nuclear device from a state inventory. But, given the extent of non-state actors' reach and capability, their nexus with state actors possessing nuclear weapons, and their control over conflict-ridden unstable states, one is compelled to revisit old assumptions on terrorists.

17. David Krieger, "The Greatest Immediate Danger to Humanity", Nuclear Age Peace Foundation, July 19, 2007.

Ideologically, the Islamic State envisions a final war with "crusader" forces for which powerful weapons are needed. It has produced and used its own chemical weapons; if it so desires, possession of nuclear weapons capability would not be out of its reach.

Meanwhile, it is also reassuring to see enhanced global efforts and unity today, thanks to the Nuclear Security Summit (NSS), to secure nuclear technology and material. An array of global as well as domestic legal provisions is in place today to curb the chances of misuse of nuclear technology. However, upkeep of nuclear weapons remains the exclusive domain of sovereign states. States follow widely varying security systems to secure their arsenals; no global rules specify how to secure nuclear weapons or the materials to make them impenetrable. Even the NSS process left the issue of security of nuclear weapons to the respective possessors. Given the changing strategic scenario, particularly in Asia, the rationality of state behaviour cannot be ensured when rogue elements control the state apparatus. If rogue elements penetrate or control state power in a country possessing nuclear weapons, and leaders are the bent on maintaining power, conscious state transfer of nuclear weapons or bomb material to terrorist groups is not unthinkable. Even high-level "rogues" within states may exploit loopholes in the system to transfer nuclear weapons, designs or material to other states—an A.Q. Khan-type clandestine network.

Furthermore, if political instability persists in a nuclear-capable country, leading to state collapse, the probability of extremist factions being able to lay their hands on nuclear material is high. As the turmoil in Pakistan continues, the safety of its nuclear arsenal is a major concern. Pakistan has a small, heavily guarded stockpile but immense threats exist—potentially outsider attacks, corrupt insiders, some with *jihadist* sympathies are a grave danger to Pakistan's nuclear assets. Other "new" nuclear-armed states perhaps represent the most alarming possibilities for nuclear diversion, particularly in the light of regional instability like the collapse of the Democratic People's Republic of Korea (DPRK).

Within this threat scenario, the list of instances of serious nuclear security and control incidents, failures, and lapses in nuclear weapons possessing countries, is long. Henry D. Sokolski and Bruno Tertrais in their edited volume *Nuclear Weapons Security Crises: What Does History Teach?* list out more than two dozen such incidents, starting from the 1960s to 2010, where nuclear weapons or technologies could have fallen into the wrong hands.[18] Some of the most egregious include: the 1961 coup in Algeria when a French nuclear site and a nuclear device were at the centre of a battle for competing loyalties; the internal power struggle within the Chinese nuclear research and development programme and within the Chinese strategic missile forces during the Cultural Revolution in 1966; the storming by anti-Moscow rebels of an army base with nuclear weapons in Azerbaijan in 1990; in July 2016, during a coup attempt in Turkey, the Turkish commander of the Incirlik Air Base, at which an estimated 50 B61 nuclear bombs stationed by the North Atlantic Treaty Organisation (NATO), gave orders to allow the anti-government forces to use the base for refuelling F-16s that bombed the Turkish Parliament.

Most incidents suggest that "political turmoil, government instability, and crisis situations put the security and control of nuclear weapons at risk of falling into the wrong hands."[19] Out of nine states that built nuclear weapons, four are known to have undergone severe political and institutional crises (Pakistan in 1958, 1977, and 1999; France in 1958, 1961; the Soviet Union in 1991, Russia in 1993; and China in 1966-68[20]) affecting nuclear security and/or control of use in one way or another.

## Increasing Global Stockpile

As long as nuclear weapons exist, and nuclear technology, material and knowhow are available, the fear and chances of their misappropriation

18. Ibid., pp. 7-12.
19. Elena K. Sokova, "Non-state Actors and Nuclear Weapons", in John Borrie et al., eds., *Understanding Nuclear Weapon Risks* (UNIDIR, 2017), p. 85. https://www.unidir.org/files/publications/pdfs/understanding-nuclear-weapon-risks-en-676.pdf.
20. Henry D. Sokolski and Bruno Tertrais, *Nuclear Weapons Security Crises: What Does History Teach?* (Strategic Studies Institute and US Army War College Press, July 2013), p. 12.

will remain. The global security discourse in vogue, and, national security strategies of the major powers still revolve around 'nuclear deterrence', but the eventual elimination of nuclear weapons will remain a pipedream for the foreseeable future. Nuclear weapons are here to stay and the world has to deal with threats to nuclear weapons.

During the two decades since the end of the Cold War, the global inventory of nuclear weapons reduced to around 13,000 today from around 80,000 during the height of the Cold War. But during the last one decade, while some nuclear weapon states, especially, China, Pakistan, UK have been in the process of expanding their inventories, other major powers like the USA and Russia have been in the process of modernising their arsenals. More importantly, miniaturisation of warheads, known as Tactical Nuclear Weapons (TNWs) that are believed to lower the nuclear threshold, are still stationed in Europe (allies of the USA) for which Sam Nunn, former US senator, made a plea for action to improve their security as they are a "terrorist's dream".[21] In South Asia, Pakistan's nuclear brinkmanship, along with its "safe haven to agents of chaos, violence, and terror" make its own TNWs "more susceptible to terrorist theft."[22]

Besides, around 520 tons of separated plutonium and 1,335 tons of Highly Enriched Uranium (HEU) stocks are located in hundreds of buildings, and bunkers, in two dozen countries without specific global rules and arrangements to secure nuclear weapons or the materials.[23]

21. Sam Nunn, "NATO, Nuclear Security and the Terrorist Threat", *The New York Times*, November 16, 2010. https://www.nytimes.com/2010/11/17/opinion/17iht-ednunn.html?_r=2&ref=global
22. "US Worried Pakistan's Nuclear Weapons could Land up in Terrorists' Hands: Official", *The Economic Times*, August 25, 2017. https://economictimes.indiatimes.com/news/defence/us-worried-pakistans-nuclear-weapons-could-land-up-in-terrorists-hands-official/articleshow/60220358.cms?from=mdr
23. Mathew Bunn, "The Evolving Threat to Nuclear and Radiological Transports". https://www.belfercenter.org/sites/default/files/files/publication/transport-threat-brief-2021.pdf

Theft of 0.01 per cent of the world's stockpile, opines Mathew Bunn, could cause a global catastrophe.

## Emerging and Disruptive Technology

The threat to nuclear assets is evolving and dynamic. It can emanate from anywhere. Besides physical threats, terrorists can use emerging technologies like Artificial Intelligence (AI), cyber attack, deep fake technology to manipulate Nuclear Command, Control, and Communication (NC3), nuclear doctrine, posturing and signalling; creating misperceptions and, consequently escalation of nuclear conflict. In times of crisis, non-state actors can create misperceptions and escalation of the conflict by generating fake videos of a leader suggesting that they have deployed nuclear weapons against an adversary. Though such attempts can be quickly detected, but once these videos become viral in social media, they will create widespread public chaos. "As 'deep fake' technology matures, it is likely to be salient in military information operations and can also create compulsions of a counterattack based on lies and fabrications."[24]

With the corruption and poisoning of data, non-state actors could stage escalatory attacks that can draw adversarial states into a nuclear crisis or falsely blame one state for the actions of a non-state actor. In December 2016, a fake news story on a site claimed that Israel had threatened to attack Pakistan with nuclear weapons if Islamabad interfered in Syria. This triggered a Twitter response none other than by Pakistani Defence Minister Khawaja Muhammad Asif, who indicated that Israel should remember that Pakistan is a nuclear-armed state. A prompt response from the Israeli Defence Ministry on the "fictitious" story put an end to the controversy. "This episode, however, highlights the potentially serious consequences of deliberate misinformation or manipulation of information, particularly in a crisis and in instances

24. Sylvia Mishra, "Deep Fakes: The Next Digital Weapon with Worrying Implications for Nuclear Policy". https://www.europeanleadershipnetwork.org/commentary/deep-fakes-the-next-digital-weapon-with-worrying-implications-for-nuclear-policy/. Accessed on November 3, 2021.

where the sides do not have reliable communication channels, or have a very short time to act on such information, or where local commanders have been delegated the authority to launch theatre nuclear weapons or submarine-based systems. Such scenarios are particularly relevant in conflict-prone South Asia."[25]

## Never Say 'No'

Unfortunately, in most instances, many, in the military and security establishment as well as intelligentsia, downplay the seriousness of nuclear terrorism. They often advance their "solemn word" that "our nuclear weapons are safe, secure and under complete institutional and professional control."[26] The UN Report *A More Secure World: Our Shared Responsibility* (2004) observes that "we live in a world of new and evolving threats where new aspects of the threat…including the rise of a global terrorist network, and the potential for terrorist use of nuclear (biological or chemical) weapons—require new responses" because "the consequences of allowing latent threats to become manifest, or of allowing existing threats to spread, are simply too severe."[27]

Unfortunately, as long as nuclear material, technology and terrorism exist, the probability that terrorists could get, and use, a nuclear bomb can never be reduced to zero. But the good part is that nuclear terrorism is both preventable and solvable. The first and final step is to acknowledge unanimously that the terrorist threat to nuclear weapons exists; the rest will follow inevitably.

25. Sokova, n. 19, p. 90.
26. "A Conversation with Gen. Khalid Kidwai". https://carnegieendowment.org/files/03-230315carnegieKIDWAI.pdf. March 23, 2015, p. 5.
27. United Nations, "A More Secure World: Our Shared Responsibility". https://www.un.org/en/events/pastevents/pdfs/secure_world_exec_summary.pdf, 2004, pp. 2-5.

# Nuclear Disarmament

# 12. Understanding Deterrence Breakdown in the Indian Context

*Prakash Menon*

## Introduction

The journey of relationships between political communities is characterised by the attempts to resolve disagreements and disputes. While the contestants wrestle for influence, coercion and persuasion are employed as tools with the aim of achieving political objectives. The spectrum of coercion and persuasion spans between extreme violence at one end, to its opposite, where violence is suspended and dialogue is the primary means of conflict resolution, at the other. Historically, coercion through threat or application of violence in its varied forms has been a frequent visitor in the strategic landscape.

Prevention of wars relies on deterrence or capitulation. In particular, deterrence has always been in the embrace of communities that were distrustful and fearful of others. It is normally driven by fear born from the imagination of loss of self-interests. The objective of deterrence is prevention of coercion, especially through violence. In operational terms, deterrence is sought to keep violence in check by using the threat of retaliation or denying its successful application through defence. In deterrence breakdown, the primacy of violence in the relationship means a state of war. Force, particularly physical force, is the tool for coercion.

As long as force is the frontrunner as a tool of choice, the nature of wars remains the same—the use of violence for political purposes. Though prolonged wars witness relative dilution of the primacy of violence, it is complementary with other means of non-violent coercion through economic, diplomatic, technological, subversive, and informational tools of statecraft.

Peace does not mean the absence of disagreements, but mostly reflects the absence of violence and the predominance of dialogue. The spectrum of struggle in relationships ranges between dialogue and coercion, based on violence. It could range from the use of indirect force coupled with deniability to full scale and direct exchange of force. Wars, on the other hand, involve the threat and use of organised violence for political purposes. They are usually waged for long, through different forms of warfare, before they reach a crescendo, and wane or persist— sometimes interminably at different temperatures.

Conflicts born of disagreements provide the seeds for war that are grown in the soil of fear, interests and honour. Conflicts are also seen as the soil that can serve nationalism, which depends, for its strength, on self-interests.

For the last seventy-five years, the advent of nuclear weapons seems to have prevented the 'big wars' of the 19th and 20th centuries. Even today, there is global political recognition that such wars should not be fought. But such recognition does not seem to be reflected in the ongoing and unbridled arms race that received its initial boost when Al-Qaeda, attacked the Twin Towers in 2001.

The international system that has evolved after World War II seems to be badly in need of repair. Healthy competition has mutated to confrontation and is driven by antagonisms born of a power struggle. The behaviour between states oscillates from being openly conflictual that sometimes involves application of force in contradistinction to a process that eschews violence. There is a growing proclivity to develop and employ the conflictual tools of national power, rather than resolve issues through a dialogue.

The instruments and tools are getting faster, sharper, accurate and deniable. If the use of cyberspace is any indication, the world has been increasingly conflictual since the end of the first decade of the 20th century. War clouds gather in the proverbial global sky, although their relative menace is seemingly greater in some regions. This essay seeks to examine deterrence breakdown in the Indian nuclear context. The natural starting point is the theory of nuclear deterrence.

## Nuclear Deterrence

Deterrence is the action of discouraging an action or event through instilling doubt or fear of consequences. It is considered a preferred option rather than dealing with the fallout of the act. Warning signs like 'Beware of Dogs', 'Trespassers will be Prosecuted' are further examples of the common practice of deterrence. The attempt is to communicate the deterrent message to prevent an act from being committed.

The three premises of the theory of deterrence are possession of adequate capability, credibility of the threat, and ability to communicate it to the adversary. Strategically, this means compelling the adversary, when faced with a given situation, to act or react in the light of the existence of a set of conditions which constitute an effective threat.

The effect, which it desires to achieve, is, therefore, a psychological one, and it is pursued by means of a threat. Inducing fear, and, therefore, inaction is the objective. The way in which deterrence acts is basically the exact opposite of war, in that the objective of deterrence is to prevent the enemy from taking a decision to act, whereas the objective of war is rather to force him to take a decision to accept the conditions imposed on him.[1]

The advent of nuclear weapons, with their awesome power of destruction, did not imbue deterrence with a newer meaning. Instead, deterrence was challenged in its accomplishment if both or all adversaries possessed nuclear weapons and the inability to defend against missiles carrying nuclear weapons. While, on the one hand, significant destruction could be inflicted on the opponent, on the other, one may also suffer large-scale destruction. Even one Hiroshima-type bomb on a major city would entail destruction that would not be politically acceptable. In the pre-nuclear era, the failure of deterrence meant an armed clash in the battlefield, resulting in victory, defeat or stalemate. The losses and gains

1. Andre Beaufre, *Deterrence and Strategy* (New York: Praeger Publications 1966), p. 24.

suffered served as a determinant of the outcome. Defence and offensive capabilities served both to limit losses to oneself and inflict them upon the opponent. Nuclear weapons changed this conventional paradigm by making victory through force application politically unviable in the context of nuclear-armed adversaries.

The theory of nuclear deterrence, therefore, relied on the ability of the defender to retaliate with nuclear weapons in spite of having been struck first. It was a capability termed as second-strike capability. Deterrence could obviously work as long as rationality prevailed on both sides. But if deterrence based on massive retaliation were to fail, it would not always be rational to carry out the stated threat in a situation of Mutually Assured Destruction (MAD).[2] This was possible because the primary aim of the retaliatory threat was to deter. But if, for some reason, deterrence failed, retaliation would serve little rational purpose, as damage would have already occurred and retaliation would only serve to invite further damage. This internal contradiction, accorded the status of a genuine paradox, undermined the theory. But by invoking 'the threat that leaves something to chance', arguing that a potential aggressor could not count on pure reason to dictate its victim's response, the theory disposed of the conundrum. Rational decision-makers realise that the laws of reason are unlikely to be obeyed strictly in a nuclear war. Even if circumstances logically compelled a rational victim of nuclear aggression to withhold a retaliatory strike, an aggressor would run the risk of inducing spasmodic retribution.[3] Also, the damage inflicted on the victim's command and control system by the aggressor's attack might, in fact, preclude any chance of rational behaviour.[4]

Rationality that was expected to underpin the calculations of costs and benefits for deterrence to prevail was unlikely to be available when it was needed most: during a crisis. The potential consequences of

2. John Steinbruner, "Beyond Rational Deterrence: The Struggle for New Conceptions", *World Politics*, Vol. 28, 1976, pp-231-32. Quoted in Bruce G. Blair, *Logic of Accidental Nuclear War* (Washington: Brookings, 1993), p. 5.
3. Prakash Menon, *The Strategy Trap* (Wisdom Tree, 2018), pp. 6-7.
4. Blair, n. 2, p. 5.

deterrence breakdown would weigh heavily on decision-makers. It could be imagined as an existential threat, regime survival or several other forms of catastrophic outcomes. In addition, psychological biases like confirmation bias and group think could result in miscommunication, misperception and misjudgment. Depending on the stage of the crisis, nuclear weapons would be on alert and the leadership would have retreated to underground nuclear shelters or be airborne. They would be bombarded by intelligence and other inputs from a plethora of sources. Assessing the situation under such condition would be a Herculean task that would also have to deal with extremely short decision-making cycles. The psychological pressure would be so great that to expect rationality in decision-making would be unrealistic. In contrast, for deterrence to prevail, rational decision-making is a necessary condition. Such a condition is unlikely to be obtained if adversaries perceive of a possibility, even if remote, that they might be visited by a few nuclear weapons.

In a crisis, rationality, the cornerstone of deterrence, is weakened when it is needed most. Deterrence is then at the mercy of fear. Nuclear strategy, therefore, has been unable to provide an answer to the question of what happens after the first nuclear weapon is fired? Because, it is impossible to know how the adversary will react. During the Cold War, by the 1980s, the idea of launching a massive 'bolt from the blue' attack to neutralise the nuclear weapons of the adversary was undermined by the large numbers of nuclear weapons in the possession of both sides and the scientific evidence regarding a 'nuclear winter' that would even pose an existential threat to humanity.[5] Operationally, the fear of a surprise attack, translated to weapons being placed on a high degree of alert even when the political relationships were on an even keel and the prospects of war were absent. This legacy of the Cold War has survived and is manifest in the alert levels being maintained by the USA and Russia even today.

5. R. P. Turco, O. B. Toon, T. P. Ackerman, J. B. Pollack, and Carl Sagan, "Nuclear Winter: Global Consequences of Multiple Nuclear Explosions," *Science*, 222, No. 4630, December 23, 1983, pp. 1283-1292. https://doi.org/10.1126/science.222.4630.1283.

There are indications that China which had so far not adopted such alert levels, may be gravitating towards keeping some of its arsenals at higher alert. The recent discovery of it building some missile silos has given birth to the suspicion.[6]

## Alert Levels

Alert levels as an operational choice are driven by the belief systems that underpin the perceived role of nuclear weapons. If perceived as having military utility, alert nuclear weapons comprise a natural choice. Striking first certainly would have advantages if one could carry out massive blows that wiped out the capability of retaliation. But as explained earlier, the long-term effects of nuclear explosions would result in a 'nuclear winter' and would tantamount to 'committing suicide for the fear of death' for the perpetrator. Scientific evidence exists that even an exchange of a few low yield weapons between India and Pakistan would have a severe long-term global impact on the Earth's biosphere, especially on its agricultural and water systems.[7]

Nuclear strategy has sought to resolve the massive exchange problem, by searching for options that, in the case of deterrence failure, can limit the nuclear exchange to a few low yield weapons. The idea being to start the nuclear exchange as signals that can deter the adversary from escalating the war and confine the exchange to within tolerable limits. The conventional threat posed by the Soviet Union to Western Europe was sought to be deterred by the threat of low yield weapons called Tactical Nuclear Weapons (TNWs). The operational concept was called a 'flexible response'. The concept certainly provided more options in theory. In practice, it has remained untested and is unlikely to successfully scale

6. Hans M. Kristensen and Matt Korda. "Nuclear Notebook: Chinese Nuclear Forces, 2021," *The Bulletin of the Atomic Scientists*, n.d. https://thebulletin.org/premium/2021-11/nuclear-notebook-chinese-nuclear-forces-2021/.
7. Ira Helfand, "Nuclear Famine: Climate Effects of Regional Nuclear War," n.d. https://www.ippnw.org/programs/nuclear-weapons-abolition/nuclear-famine-climate-effects-of-regional-nuclear-war#:~:text=The second edition of IPPNW's, global damage to the earth.

the firewall of potential escalation. To understand the problem, one has to turn to the explanation of escalation as enunciated by Clausewitz, termed the trinity.

**Fig. 1: The Trinity**

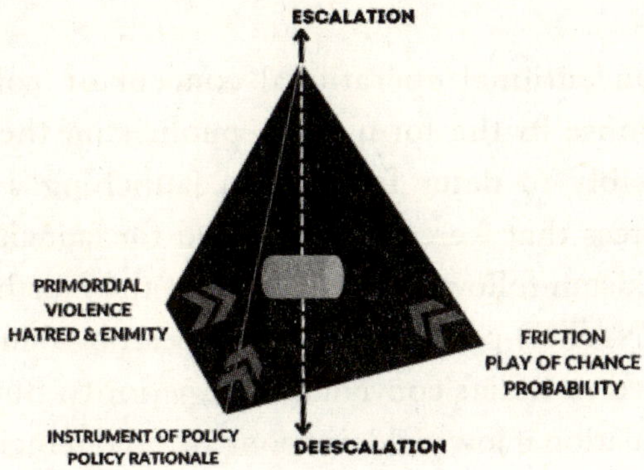


Created by Prasanna Naidu

Each state has its own trinity that is in tension with itself and is the product of the interaction with the opposing side. Escalation and deescalation are determined by the strength of three elements that generate forces that exert themselves on the magnet suspended between them. Pulls or pushes correspond to either escalation or deescalation. Illustratively, the strongest force in the trinity that could prevail in the aftermath of terrorist strikes on India would be the elements of hatred and enmity between the societies of India and Pakistan. The danger lies in the possibility that the forces of enmity and hatred could overcome the element of political rationality that is expected to deescalate the crisis. However, what is of major concern is the combination of hatred and enmity and the uncontrollable elements of friction, play of chance and probability which can, at the extreme, bring nuclear weapons onto alert. Once on alert, the probability of accidental and inadvertent nuclear exchange increases and harbours the potential for a breakdown of nuclear deterrence.

Once military forces are engaged at the conventional level, their interaction could acquire a logic of its own that may not be amenable to control by political rationale. This could be overwhelmed by the deep sense of distrust and inability to judge the intentions of the adversary. In any case, notions like 'flexible response' have not been battle tested, though they remain in the doctrinal cupboards of the USA, Russia and Pakistan.

India's conventional operational concept of 'cold start' evoked Pakistan's response in the form of its publicising the deployment of TNWs, ostensibly to deter India from launching a speedy ground offensive by forces that were pre-positioned for launch at short notice. Apparently, Pakistan followed the reaction of the North Atlantic Treaty Organisation (NATO) powers in the European battlefield to neutralise what it perceived as India's conventional superiority. But the problem of controlling escalation if low yield weapons are used remains an intractable one and entails risks that could have catastrophic consequences.

The issue of deterrence breakdown in the Indian context requires to be viewed primarily in the Sino-Indian and Indo-Pak contexts. Three primary forces are believed to have significant roles in the breakdown: political and military misjudgment, alert levels and escalation potential.

## Sino-Indian Context

In geographic terms, the Sino-India confrontation is likely to remain in the continental and maritime realms and represented primarily by the Himalayan border and the Indian Ocean. The disputed Himalayan border was home to the 1962 War that was preceded, and followed by, several military incidents which involved loss of lives and some changes in territorial status quo. However, for nearly three decades, during the period of the early Eighties to nearly the end of the first decade of the 21st century, relative calm prevailed. From 2011 onwards, frequent but small scale military incidents had increased in frequency and reached a peak in April 2020, when the People's Liberation Army (PLA), occupied several disputed areas in Ladakh. The Galwan incident of June 2020, resulted in casualties to both sides and it happened during the process of talks that was underway between the military commanders.

Though broad political level understanding has been reached to de-escalate and resolve the territorial disputes, both sides have mobilised and continue their operational preparations, which now span the entire Himalayan border. Though deescalation, in terms of proximate deployment, has been achieved in Ladakh, the continuation of the status quo favours China. As of this writing, both sides seem likely to stay the course and face the harsh Himalayan winter.

The opening up of another front by China in the Himalayas has added to the hotspots, where it was already engaged. These include Taiwan, Senkaku Islands and its claims in the South China Sea. There has been a lot of speculation regarding China's military moves in Ladakh and its political intentions. For sure, it is related to the ongoing higher level geopolitical struggle with the USA. Therefore, a plausible explanation for China's Ladakh venture could be related to keeping India confined to the subcontinent. Strategically, getting India to expend its military resources for the defence of its northern borders could slow down India's development of its maritime power. Combined with using Pakistan as a south paw, which it has been doing for long, the impact on India is maximised. The operational methodology has conformed to salami slicing which has been its signature tune in the other hotspots of Taiwan, East Asia and the South China Sea.

In the ongoing Ladakh crisis, China violated long-standing agreements that had preserved an uneasy peace. Previously, both countries used to patrol the disputed areas and return to their bases. In April-May of 2020, China established physical control of some of these areas through military occupation. The occupation process did not involve any physical eviction of Indian troops, except in the Galwan clash.

The extant military situation holds potential for a larger confrontation and conflict if China shifts from salami slicing to launching a military offensive. Any large scale military offensive in sensitive areas like Depsang in Ladakh or the Siliguri Corridor in the northeast would evoke an Indian reaction that could escalate into war. However, if China's intentions are confined to keeping India engaged on the northern border, it is unlikely to risk being tied down to a prolonged conflict when its priorities lie elsewhere in Taiwan and the South China Sea. Though it could succeed in sucking up India's meagre resources that have also taken a beating due to COVID-19, the impact in the context of the higher geopolitical confrontation may work against its interests. This is so because India has shifted its geopolitical weight in the maritime domain by seeking to deepen its cooperation with the USA and other powers like Japan, France and the United Kingdom. Such a move was what China was seeking to prevent, for its vulnerabilities lie in the protection of trade routes in the maritime domain, especially in the Indian Ocean region where India is geographically advantaged. China continues to seek bases in the Indian Ocean region and has already established some at Gwadar in Pakistan and Djibouti in the Horn of Africa.

The probability of a direct military confrontation in the Indian Ocean region remains low, especially since both countries are dependent on the same trade routes. Paradoxically, despite the ongoing Ladakh crisis, the quantum of trade between both countries has increased. Therefore, unless a major war breaks out in the continental space of the Himalayas, it is difficult to visualise a situation where any Sino-Indian military confrontation can ensue in the maritime domain. Instead, what is likely to be witnessed is the juggling for influence that involves India's neighbours and the countries of the Indian Ocean littoral.

A major war in the Himalayas that threatens India's sensitive spots, mentioned earlier, could be imagined as a result of political misjudgment regarding India's political will to resist. Xi Jinping could misjudge India's resolve and seek a military victory that serves to signal to other powers that they too can face a similar fate. But considering the ongoing internal and external challenges it faces, China's strategic landscape is getting difficult as its continued aggressiveness has resulted in several powers coalescing to put up a common front.

Nuclear weapons can be expected to prevent a major Sino-Indian conflict. Any conflict in which nuclear weapons are alerted should be a major cause for concern. This is despite the fact that both countries propound no first use, but once nuclear weapons are alerted and engage interactively through deployments, the probability of nuclear use increases, not due to a deliberate decision taken of nuclear use but because nuclear deterrence can break down accidentally as a result of the forces illustrated in Clausewitz's trinity. The possibility may be remote, but it should be sufficient to induce caution in political leaders. Overall, the impact of nuclear weapons may confine itself and follow the stability-instability paradox. Low levels of conflict under the nuclear shadow could precariously keep nuclear deterrence from breaking down.

## Indo-Pak Context

India and Pakistan had fought three wars prior to both nations going nuclear in 1998. The Kargil conflict that followed in 1999 was fought when both nations possessed rudimentary nuclear capability. Historical evidence of Pakistan's growing involvement in Jammu and Kashmir(J&K) suggests that its involvement growth curve trajectory paralleled its nuclear capability,[8] which at the time of Kargil consisted of a very limited but unknown number of nuclear warheads. Its delivery capability was restricted primarily to a few aircraft (F-16s or Mirages) suitably modified. Though Pakistan had earlier carried out several tests of short-range (Hatf and Shaheen) and medium-range (Ghauri) missiles, none was operational at the time of the Kargil War. Pakistan, at the time of the Kargil conflict, possessed an opaque and highly limited number of undeployed nuclear weapons and restricted delivery capability. The arsenal, by former President Musharraf's own admission, was not operational.[9] It was a non-existent capability around which Pakistan wove its paradigm of a nuclear umbrella, to prevent escalation as well as paint the Indo-Pak situation as being of an extremely dangerous nature to leverage the nuclear flashpoint card internationally and invite intervention.

---

8. The correspondence of the acquisition of nuclear capability and the prosecution of war is illustrated in Jasjit Singh, "Kargil 1999: Pakistan's Fourth War of Kashmir", *Strategic Analysis*, Vol. 13, No. 5, IDSA, August 1999, p. 691.
9. Pervez Musharraf, *In the Line of Fire* (UK: Simon & Schuster 2006), pp. 97-98.

India weaponised its nuclear capability between 1992 and 1994.[10] The development had been underway since the late 1980s and a former Chief of the Air Staff (COAS) told the Kargil Review Committee that he was inducted to work for the nuclear delivery programme some time in 1986. It is, therefore, possible that Pakistani threats between 1984 and 1990 were founded on the notion of nuclear asymmetry, a topic that was being widely debated in India during that period.[11] Ever since the demonstration and declaration of nuclear capability by India and Pakistan in May 1998, the overt situation had changed. Though both sides possessed only rudimentary arsenals, the mere possibility of suffering destruction on a catastrophic scale was sufficient as deterrence.

The Lahore Declaration issued at the end of the meeting between Prime Minister Atal Bihari Vajpayee and his Pakistani counterpart Nawaz Sharif in early 1999 affirmed that "the nuclear dimension of the security of the two countries adds to their responsibility for avoidance of conflict between the two countries."[12] After the 2001-02 crisis, India propagated the concept of limited war which does not quite fit into the Pakistani paradigm of the nuclear umbrella facilitating a proxy war without inviting an Indian response.[13]

Speaking at a seminar in November 2002, General VP Malik elaborated on the concept of limited war:

> The fundamental point for a limited war is that it is a political process conducted for bargaining. The aim is not to win but rather not to lose, and fight in such a way that the enemy is forced to settle for peace....

10. Ibid., p. 205.
11. K. Sundarji, *Blind Men of Hindoostan* (New Delhi: UBS Publishers Distributors, 1993). In a mix of fact and fiction, Sundarji demonstrated the folly of nuclear asymmetry in the Indo-Pak context. See also, "Effects of Nuclear Asymmetry on Conventional Deterrence", *Combat Papers*, No. 1, College of Combat, Mhow, May 21, 1981. In contrast, the present debate is focussed on application of conventional power in a situation of nuclear parity.
12. Memorandum of Understanding, signed by the Indian and Pakistan Foreign Secretaries, in Lahore, Pakistan, February 21, 1999. <http://www.indianembassy.org/south_asia/pakistan/mou (lahore01211999).html>
13. Menon, n. 3, pp. 98-99.

> There is a linkage between deterrence and escalation. Capability to wage a successful conventional and nuclear war (if necessary) is an important deterrent. A war may well remain limited because of credible deterrence....Continuous control of the escalatory ladder by the political and military leadership is important[14].

The possibility of control remains subject to Clausewitz's trinity and has possibly succeeded because of the caution displayed by the political leadership on both sides. After the 2001-02 parliamentary attack crisis, the terrorist attacks at Mumbai in 2008, Gurdaspur in 2015, and Pathankot in 2016 elicited no overt military response from India. However, after the attacks at Uri in 2016 and Pulwama in 2019, India responded militarily. Post Uri, it was through ground-based attacks on Pakistan's infrastructure that supported terrorist activities, and after Pulwama, it was through air strikes at Balakot. In both cases, Pakistan denied the destruction inflicted while India claimed success.

Pakistan, especially its military, utilised the Indian reaction to magnify the Indian threat to buttress its position in domestic politics and project the Indo-Pak situation as a nuclear flashpoint. On the other hand, India's political leadership leveraged the tough stance taken for considerable gains in electoral politics. Overall, India's publicised military reaction has not stopped Pakistan from giving up terrorism as a tool of foreign policy. With the abrogation of Article 370 and the bifurcation of the state of J&K in 2019, Indo-Pak relations further deteriorated and as of 2021, there were growing signs of Pakistan enhancing its support to terrorism in J&K. With the US withdrawal from Afghanistan in August 2021 and the Taliban ascending to power, Pakistan seems to be emboldened to deepen its support to terrorism in Kashmir. A move that could also be fanned by India's preoccupation in confronting China's threat on the northern border. The danger of deterrence breakdown has, therefore, grown—a possibility that is fertilised by the trajectory of domestic politics in both nations.

---

14. VP Malik, "Limited War" in Singh, Jasjit, ed., *Air Power and Joint Operations* (New Delhi: KW Publishers, 2003), p. 157.

## A Plausible Scenario

In India, the ideological disposition of the government has resulted in communal polarisation on a scale that has not been witnessed since the partition of 1947. The support for a Hindu majoritarian agenda enjoys widespread support, especially in the densely populated states of the Indo-Gangetic plains. The primary tailwind of electoral politics now derives its power from communal politics. In Pakistan, extremist forces have received a boost with the Taliban in power in Afghanistan and triggered a churn that has the military, the civilian government and extremist forces as its primary contestants. Pakistan's military and the extremist forces could perceive India's communally polarised ambience as an opportunity to further its strategy of 'bleeding India with a thousand cuts'. Pakistan's actions could play out according to this imagined script given below.

Pakistan launches a terrorist attack on a Hindu religious congregation somewhere in the Indo-Gangetic plains and inflicts large scale civilian casualties. The intention is to provoke an anti-Muslim pogrom. The attack would be denied by Pakistan. Having already established its adoption of having taken a tough stand against Pakistan and driven by the calls for revenge, India would react militarily. An action-reaction cycle would be set in motion that would now be at the mercy of the 'trinity'. The potential for the breakdown of deterrence at both conventional and nuclear levels now lies in the unknowable role the three forces of the trinity will play.

The red flag in any military confrontation is the alerting of nuclear weapons. Once alerted, the chances of accidental nuclear war are increased. This is so because mistrust is heightened and even defensive moves of the nuclear arsenal can be perceived as offensive, the fog of war being the breeding ground for misjudgment, miscommunication and miscalculation. The inability to differentiate between conventional and nuclear carrying aircraft and missiles is an intractable problem that is to some extent ameliorated by India's No First Use(NFU) policy. Though both countries may draw their red lines for nuclear use, the deliberate initiation of a nuclear exchange is unlikely, though not impossible.

To presume that deterrence will hold in all circumstances ignores the unknowable role that the forces in the trinity can play.

## Conclusion

Deterrence breakdown at the conventional levels in the Sino-Indian and Indo-Pak contexts will remain a possibility as long as geopolitical tensions remain unresolved and force application in any form is resorted to. The three countries are presently imprisoned in the framework of an unbridled global arms race that is driven by breakdown in relations and technological advancements that promise better defensive and offensive capabilities.

The silver lining that could keep deterrence breakdown at bay is the fact that emergency communication channels have been established between the political and military leaders of India and China, and India and Pakistan. There are also several agreements and memoranda of understanding that have been signed to prevent military escalation. The Sino-Indian agreements are presently weakened by China's actions in Ladakh. But as long as China does not switch from salami slicing to launching a major offensive, conventional and nuclear deterrence can play a major role to limit escalation. With Pakistan, as long as it continues to use terrorism as a tool of foreign policy, the possibility of a limited conventional war remains on the table. Such a possibility harbours a breakdown of conventional deterrence that, in turn, could challenge the ability of nuclear deterrence to prevent the use of nuclear weapons. The danger of nuclear deterrence breakdown then lies in uncontrollable forces that escape the grasp of political rationality.

War avoidance as a strategic goal for all parties is the way forward. But considering the salami slicing tactics of China and use of terrorism by Pakistan, the stability-instability paradox might keep the 'big wars' at bay even as it harbours the possibility of uncontrolled escalation to the nuclear realm.

The solution to the paradox does not lie in the search for building more conventional and nuclear capabilities but in the political realisation that the utility of force is limited under the nuclear shadow and cannot possibly provide relief to political disagreements. This reality is obvious but is blinded by the dynamics of domestic and international politics.

With the global geopolitical ambience being governed more by preparing for war, in order to prevent it, the chances of deterrence breakdown increase not so much because of the human agency but because of loss of control during a military confrontation.

# 13. Mapping Arms Control and CBMs: From Cold War Years to Present Times

*R. Rajaraman*

Arms control and CBMs (Confidence-Building Measures), each represents an important set of actions designed to reduce the dangers posed by nuclear weapons. But, although directed at this common goal, they are quite distinct from one another at the conceptual level. Therefore, to begin with, the chapter discusses them separately. Then, it points out the linkage between them—how transparency and verifiability in adopting arms control measures by any nation act as important CBMs with respect to its friends and foes alike.

## Arms Control

By the end of the Sixties, the number of countries with nuclear weapons had grown to include all the P5, namely, the five permanent members of the UN Security Council with veto rights—the US, USSR, UK, France and China. Even as they were building their weapons, they began to develop concerns, as self-appointed guardians of the world's security, about more countries (such as Israel, India, Pakistan) joining the nuclear club.

Thus, even as the arsenals of the US and USSR were continuing to grow into tens of thousands, they had started thinking about ways to control the rapid growth of nuclear weapons elsewhere. In fairness, they had also begun a process of capping and eventually reducing the size of their own nuclear forces.

This led to a set of measures and treaties, largely pushed by the US, which may be collectively called "Arms Control". It encompasses several different families of initiatives. Two major components of arms control are:

- Arms reduction and disarmament of arsenals of existing nuclear powers or vertical non-proliferation.
- Preventing more nations from building nuclear weapons or horizontal non-proliferation.

Of these, the second item in the list, horizontal non-proliferation, has perhaps been the most substantive nuclear arms control preoccupation of the international community—both nuclear and non-nuclear nations.

## Arms Reduction

Arms reduction has so far been largely a bilateral matter, undertaken by the US and the USSR (now Russia) after extensive negotiations of technical details. By 1969, even as the number of warheads kept growing, mercifully, the two Cold Warriors also started talks in parallel to contain the arms race. These resulted in a sequence of treaties in the next three decades, leading to concrete reduction in weapons.

First came the Strategic Arms Limitation Treaties—SALT I (during the Brezhnev-Nixon regime in 1972) and SALT II (during the Brezhnev-Carter regime in 1979). When the reductions negotiated in the SALT Treaties were implemented, the total number of warheads in the world had come down from over 60,000 at the peak of the Cold War to about 22,000 by 1991.

Then came the Strategic Arms Reduction Treaty—START I (1991), which further lowered the number of warheads by about half, in a verifiable manner. This was followed by the Moscow Treaty of 2002. By 2005, over 40,000 weapons (≈70 per cent of the peak Cold War figure) had been consigned for dismantlement.

The next major bilateral arms reduction treaty was the New START Treaty negotiated by Presidents Obama and Medvedev. Obama worked strenuously to complete the negotiations fast, and the New START

Treaty was signed at Prague by the two of them on April 8, 2010. The treaty's duration was to be ten years, with an option of extending it once for five years. The treaty set a limit of *700 deployed delivery vehicles and 1,550 warheads* for each country. This limit was 74 per cent lower than the limit of the 1991 START Treaty and 30 per cent lower than the Moscow Treaty.

Altogether such a major reduction in the superpower nuclear arsenals in a *mutually verifiable manner* was an admirable and extremely difficult diplomatic and technical achievement: a colleague I met in the US who used to work in their weapons laboratories told me that it takes roughly three days to carefully dismantle a single nuclear weapon.

The New START Treaty was due to expire in February 2021. Letting an arms reduction treaty expire is a very serious matter. Not only are the limits on nuclear arms set by that treaty no longer operational, but in the absence of any live treaty, the nations involved are free to rearm themselves back to as high a level as they wish to. But, fortunately, after some suspense during the Trump presidency, the New START Treaty has now been extended up to 2026 by the Biden Administration.

In addition to the US-Russian arms reduction, the UK and France also reduced their comparatively smaller nuclear weapons stocks even further. Of course, they are both covered by the nuclear umbrella of the North Atlantic Treaty Organisation (NATO) alliance.

In our own backyard, there have been no similar attempts between India and Pakistan to negotiate some arms reduction between them. To the best of our knowledge, gleaned from publicly available information, both India and Pakistan are continuing to produce nuclear weapons—or, to be more precise, the fissile materials need to fuel them.[1] (There is no public information on how much of the South Asian stock of fissile materials has been weaponised).

1. For full available information of fissile material production in South Asia and elsewhere, see the Global Fissile Material Reports of the International Panel on Fissile Materials (IPFM). www.fissilematerials.org

## Disarmament

Arms reduction, even by individual (or pairs of) countries, has itself been very painstaking, with much preparatory work on the diplomatic and technical fronts. Disarmament, by which one usually means total and global elimination of nuclear weapons, is clearly far more difficult to achieve. Nevertheless, there have been a few genuine attempts over the years.

In fact, calls for nuclear disarmament are as old as the nuclear era itself. Inspired by intellectual giants like Bertrand Russell and Albert Einstein, the Pugwash Movement was born, dedicated to eradicating nuclear weapons. Proposals for disarmament were also offered by nations led by statesmanlike individuals. India was among the first countries to passionately argue in the UN for disarmament.

Many years later, Prime Minister Rajiv Gandhi of India took several initiatives towards universal disarmament. He presented on June 9, 1988, a time-bound action plan for achieving a nuclear weapon-free world at the third Special Session on Nuclear Disarmament at the United Nations. He was also part of the Six-Nation Initiative, which was founded in 1984 to lobby for disarmament. Its members were Sweden, Greece, India, Tanzania, Argentina. The leaders of these six countries called on the United Nations to create a multinational system to verify arms control agreements and put in place an orderly process of universal disarmament. Unfortunately, even though they had a serious well-argued action plan for total disarmament, it was not taken up seriously by the major nuclear powers. This was to be expected. As one can imagine, demands for disarmament from countries that had no weapons themselves will have little effect in *realpolitik*.

One would have hoped that real progress on disarmament would happen when the nuclear superpowers themselves, in their own wisdom, chose to reduce their stockpiles. It was, therefore, disappointing when efforts towards eliminating nuclear weapons even by the leader of the most powerful nuclear power in the world also came to naught. We are referring to Mr. Barack Obama, who, when he was running for the US presidency in 2008, included nuclear disarmament as a major plank of his platform—the first US presidential candidate to do so. And he walked the talk after being elected, becoming the first US president in history to openly call for ridding the world of nuclear weapons. He was an outstanding orator and gave a series of inspiring speeches, which were widely admired around the world, and his initiative for nuclear disarmament welcomed, albeit somewhat hypocritically, by political leaders all over. In fact, he was awarded the Nobel Peace Prize for his efforts.[2]

In the event, he could not mobilise enough support even among US Congressmen for total disarmament. Just the process of pushing the bilateral US-Russia New START Treaty through the US Congress required much political persuasion and cashing in of goodwill from senators. So much so that another of Obama's arms control goals, namely, getting the US Senate to ratify the Comprehensive Test Ban Treaty (CTBT), already signed by the US, had to be abandoned. Getting acceptance of total disarmament was an unrealistic goal. Thus, sadly, global disarmament continues to be a utopian dream, despite efforts by President Obama and Non-Governmental Organisations (NGOs) like Global Zero.

## Non-Proliferation

The non-proliferation component of arms control has been institutionalised through the coming into force of the Treaty on Non-Proliferation of Nuclear Weapons (NPT) in 1968.

From the time the nuclear era dawned in the mid-Forties, nuclear weapons were considered, leaving aside the humanitarian and moral issues surrounding them, extremely complex and delicately engineered constructs, requiring a support base of industrial and scientific sophistication. To be able to develop these weapons and successfully test them, was viewed as an indicator of technical advancement. On top of that, the fact that by 1964, the only countries to possess their own nuclear

2. In my opinion, and that of many others, the award was premature and his eager acceptance of it was a disappointing blot on his record.

arsenals, namely, the US, USSR, UK, France and China, were also the five permanent members of the UN Security Council reinforced the notion that nuclear weapons were not just instruments of mass destruction but also a currency of global power and prestige. Inevitably, this added to the motivation of nations that were already nuclear powers to hold on to their special status, in addition to genuine concerns about the dangers of nuclear proliferation. As a result, once they all had nukes in their pockets, the P5, as trustees of the world's security, decided that it was too dangerous to allow any more countries to make them. They were able to push through resolutions in the UN to enable the birth of the NPT in 1968.

The NPT divided the world into the haves and the have-nots:
(i) Nuclear Weapon States (NWS), i.e. the P5; and
(ii) Non-Nuclear Weapon States (NNWS).

The treaty disallowed the NNWS from acquiring the weapons, and brought in instruments of verification through safeguards implemented by the International Atomic Energy Agency (IAEA).

Highly asymmetrical though this arrangement was, the NPT was amazingly successful in getting 190 signatory state-parties. Among the carrots offered to the NNWS were:
(a) help in developing civilian nuclear power; and
(b) the promise by the NWS (in Article VI of the NPT) to "pursue negotiations....on effective measures relating ... to nuclear disarmament, and on a treaty on general and complete disarmament".

There were other factors that also made the NPT acceptable to the NNWS. Most of them were, in any case, developing nations without the technological base, or the economic strength to go in for nuclear weapons. Of the developed countries among the NNWS, a few renounced the nuclear option on principle. But a large fraction of the NNWS in Europe were already covered by the nuclear umbrellas of the US and USSR.

Only India, Pakistan and Israel did not join the NPT. North Korea withdrew in 2003. [Note: Despite the current controversy over its having produced Highly Enriched Uranium (HEU), Iran continues to be a part of the NPT].

### More Arms Control Acronyms: FMCT and CTBT

Major offshoots of the NPT regime were the efforts to establish two other treaties, namely,

1. The Fissile Materials Cut-off (Control) Treaty (FMCT); and
2. The Comprehensive Test Ban Treaty (CTBT)

Let us begin with the FMCT. Fissile Materials (FM) are those that generate nuclear energy. In the case of the more familiar carbon-based fuels such as coal, oil, explosives like TNT, etc., their energy generation comes only from chemical reactions, which involve the periphery of atoms, leaving the nuclei inside those atoms essentially untouched. By contrast, fissile materials, by having their atomic nuclei break up through the process of nuclear fission, generate prodigious amounts of energy. Consequently, they are used as fuel for both nuclear reactors and weapons. The most widely used fissile materials are weapon grade Plutonium (Pu) and Highly Enriched Uranium (HEU).

Given that FM are key constituents of nuclear weapons, a substantial component of the international effort towards nuclear arms control and global disarmament has been prevention of the proliferation of these materials, capping their further production and, eventually, eliminating them. In the meantime, it is also important to maintain a very tight inventory of these materials, wherever they are, and ensure their security so that they don't fall into the "wrong" hands such as terrorists.

Such vigilance is exceedingly difficult to maintain since it takes pilferage of just a very small fraction of the stock of fissile materials available around the world to make a nuclear bomb. Typically, you need about 20-40 kg of HEU to make a weapon. Plutonium-fuelled bombs require even less—about 5-7 kg of weapon grade plutonium will suffice. Compared to this, the total amount of these materials produced and stocked by different nuclear weapon powers is huge. The best available estimates are to be found in the reports of the International Panel on

Fissile Materials (IPFM), a body that studies various different aspects of fissile materials and keeps track of their stocks in different countries.

According to the latest figures in the IPFM's website[3], there are about 1,330 tons of HEU and 540 tons of separated plutonium in the world. At median values of,say, 6 kg of Pu or 30 kg of HEU to make one weapon, these stocks are sufficient to make 90,000 Pu-based weapons and 44,000 HEU-based weapons respectively! Together, the roughly 1,34,000 nuclear warheads that could in principle be made out of the total stock of FMs is *more than twice* the combined arsenals of both the US and the USSR during the peak of the Cold War, which added up to only (*sic!*) about 60,000 warheads. Clearly, efforts to disarm the existing nuclear arsenals will be ultimately fruitless if the process may be easily reversed by the availability of so much fissile materials.

These concerns led the nuclear powers to cap fissile material production and stocks.

The P5 nations, namely, the US, USSR (now Russia), France, UK and China, were also the first to start producing FM and assembling them into nuclear weapons. Among them, *all except China* have announced a moratorium on further production of weapon-grade FM. They are believed to have already produced much more FM than they need for making nuclear warheads. China has not formally announced a moratorium but is believed to have also stopped production. Presumably, it wishes to retain the right to produce more if, in the future, its security perception demands it. Some, like the US and the UK, have also publicly announced their inventories of FM stocks. The other nuclear weapon nations, viz. China, Israel, India, Pakistan and North Korea have not made their FM holdings public. Scholars and analysts have had to estimate their stocks the hard way by using whatever information could be gathered from assorted sources and make estimates, using physics, based on the numbers, models and capacities of their reactors. Such estimates show that the Indian and Pakistani HEU holdings are miniscule compared to those of the two nuclear superpowers. The Indian holding has now risen from

3. Ibid. See their tables dated September 21, 2021.

0.2 tons to 3.2 tons since 2006, driven by its HEU production intended for its nuclear submarine reactors, whereas the additional HEU stocks of Pakistan are presumably intended for its weapons. During the past decades, Pakistan has also been producing plutonium in its four reactors at Khushab. If one goes by publicly available reports, both countries seem to be continuing to produce more fissile materials.

Given the large existing stocks of FM around the globe and the possibilities of more countries going into FM production, efforts were made to cap and control FM production globally. In 1993, a resolution was passed at the UN General Assembly calling for".....a verifiable treaty banning the production of fissile material for nuclear weapons......". To this end, the UN Conference on Disarmament (CD) in Geneva was charged with immediately commencing negotiations on a treaty to control and cap FM production. Discussions began on such a treaty popularly denoted by the acronym, FMCT where the letter 'C' could stand, depending on your political position on the scope of the treaty, for "control" or " cut-off".

However, despite much discussion at the CD in the last 20 years, actual negotiations on such a treaty could *not even get started*. Various procedural objections were raised by different member states, and conflicting drafts of the treaty document were presented by different countries, including the US and China. The IPFM also produced a draft treaty document for the FMCT, which was jointly submitted to the CD by the delegations of Japan, Canada and the Netherlands in September 2009.[4] The IPFM's draft did not represent any one country's view but was intended to provide the member states of the CD with useful resource material for our work in the prospective negotiations on a Fissile Material Cut-off Treaty.

All these efforts were in vain and none of the draft documents was fully acceptable to all the members of the CD, which operates by consensus, thereby effectively giving every participant nation a veto.

4. https://fissilematerials.org/library/2009/02/draft_fissile_material_cutoff_.html

Pakistan, in particular, wanted the scope of the treaty to include not just future production of FM but also existing stocks. Underlying its concerns were the stocks of reactor-grade plutonium that India had accumulated from its civilian power reactors. These stocks were kept out of the scope of IAEA safeguards under the terms of the Indo-US nuclear deal. Pakistan claimed that these stocks of India's reactor-grade plutonium give the latter the option of using them for further increasing its nuclear arsenal. India has always maintained that it intends to use these stocks only for civilian power generation and not for weapons, and, consequently did not support the inclusion of "existing stocks" within the purview of the proposed treaty. Inclusion of existing stocks was also not acceptable to many other countries, including China. This had stalled progress in negotiating the FMCT at the CD.

In recent years, Pakistan has dug its heels in and explicitly declared that it is opposed to any negotiations on fissile materials control. Attempts to persuade Pakistan to change its mind have not been particularly vigorous, given the continuing reliance of the US on Pakistan for other strategic purposes as well as China's support to it. For an expert discussion of Pakistan's evolving position on FMCT, see the excellent review by Zia Mian and A.H. Nayyar.[5]

Given the constraints of consensus at the CD, efforts were also made to negotiate a fissile materials agreement outside the CD, but these have not borne fruit so far. As things stand, there is no internationally agreed control or cut-off on fissile materials.

## The Comprehensive Test Ban Treaty (CTBT)

Clearly, a vital step in the development of any new device is to test it to see whether, and how well, it works. That holds for nuclear weapons too.

5. Zia Mian and A.H. Nayyar, "Playing the Nuclear Game: Pakistan and the Fissile Material Cut-off Treaty", *Arms Control Today*. https://www.armscontrol.org/act/2010-04/playing-nuclear-game-pakistan-fissile-material-cutofftreaty#:~:text=%5B1%5D%20The%20treaty%20would%20ban,work%20that%20included%20an%20FMCT.

Tests are conducted by the nuclear powers not only at the beginning of their process of building a nuclear arsenal, but also at later stages as they try to improve the design, quality and explosive yield.

Thus, before dropping its nuclear bombs on Hiroshima and Nagasaki in August 1945, the US conducted its first nuclear test on a prototype on July 16, 1945, in the deserts of the state of New Mexico. Subsequently, the US continued its tests, as did the USSR when it started building its own nuclear arsenal. Initially these tests were conducted above the ground.

But, as the radiation damage due to the aboveground tests became more apparent, the testing process was moved in part underground and in part (in the case of the US) to remote islands and knolls in the Pacific Ocean. The development of the thermonuclear weapon (popularly called the H bomb) by the US and soon thereafter by the USSR, required more and even bigger tests.

A major turning point in nuclear testing came at the time of the Cuban missile crisis during October 16-28, 1962. As is well known, that crisis was created by the Soviet Union clandestinely transferring nuclear missiles to its ally Cuba, where they were being deployed, ready to hit American targets. More such missiles were on their way to Cuba by sea. (To be fair, we must remember that the US already had its nuclear missiles stationed in countries like Turkey, close to the borders of the then USSR). Once the US detected these nuclear armed enemy missiles deployed in Cuba, right next door, as it were, not far from the US east coast, it exposed the findings to the international community and started planning an immediate and urgent response against the Soviet action. With the US and USSR facing each other eyeball-to-eyeball, the world came closer to a nuclear war than ever before. Fortunately, for mankind, within a couple of weeks of this confrontation, the USSR backed down. Its ships carrying more missiles to Cuba were stopped and the deployed missiles in Cuba were dismantled.

The Cuban missile crisis was diffused quickly , but there was a strange event that took place right in the midst of this crisis. The USSR tested a massive 300 Kiloton (kT) warhead[6] in outer space on October 22, 1962, detonated at an altitude of 290 km above Dzhezkazgan in Kazhakstan. At that altitude, the pressure, heat and radiation waves of the nuclear explosion did not cause any major structural or health damage on the ground. But nuclear explosions also emit a very strong Electro-Magnetic Pulse (EMP). In this outer space Soviet test, the EMP fused 570 km of overhead telephone line with a current of 2,500 amperes, started a fire that burned down the Karaganda power plant, and shut down 1,000 km of shallow-buried power cables between Aqmola and Almaty.

That such a major nuclear test causing so much infrastructural damage could be conducted right in the middle of the Cuban missile crisis shocked both superpowers into negotiating and agreeing to the Partial Test Ban Treaty (PTBT). The original parties to the treaty were the US, the UK, Northern Ireland, and the USSR. Each of the parties undertook to prohibit, prevent, and not carry out any nuclear weapon test explosion, or any other nuclear explosion, at any place under its jurisdiction or control:

(a) in the atmosphere, beyond its limits, including outer space; or under water, including territorial waters or high seas; or
(b) in any other environment if such explosion causes radioactive debris to be present outside the territorial limits of the state under whose jurisdiction or control such explosion is conducted, in the atmosphere, deep space or underwater.

The treaty entered into force on October 10, 1963.[7] It did not stop nuclear testing by any means, but restricted it to underground tests. The major nuclear powers continued with their tests, albeit underground, for another quarter of a century. A shocking statistic is that during the peak

6. The yield of a nuclear weapon is quoted in kilotons (kT), which stands for the equivalent amount of the conventional detonator TNT needed to produce the same explosive power as the warhead.
7. "Inventory of International Nonproliferation Organizations and Regimes", Centre for Nonproliferation Studies, Monterey.

period, a nuclear test shook and often irradiated the earth on an average every nine days.[8]

As the weapon designs got perfected, the need for further tests went down. The USSR stopped all testing in 1990, and the US in 1992. The remaining nuclear powers of that time, the UK, France and China last tested in 1991, 1996 and 1996 respectively.[9] Altogether, over 2,000 tests were done by 2017, some underground and others in the atmosphere.[10]

Meanwhile, even as they were conducting their own nuclear tests, the superpowers proceeded to repeat, in the name of non-proliferation, the same autocratic steps that they had taken in the context of the NPT and FMCT. Having more or less done all the nuclear testing they needed to do, they wanted to prevent any other country from conducting nuclear tests. They set in motion negotiations on a Comprehensive Test Ban Treaty (CTBT) that would ban all forms of nuclear explosions, including nuclear weapon tests conducted underground.

In 1996, after two and a half years of intense engagement, India walked out of the CTBT negotiations at the Conference on Disarmament, with our Ambassador, the late Arundhati Ghose famously standing up to Western pressures. Eventually, the CTBT did open for signature at New York on September 24, 1996. But it is not in force even today because it requires, as listed in an annex to the treaty, ratification by all the 44 states possessing nuclear power reactors or nuclear research reactors. Three of these states—India, North Korea and Pakistan—have neither signed nor ratified the treaty, and a further six—China, Egypt, Indonesia, Iran, Israel and the United States—have signed but not yet ratified the treaty.

India refusing to sign CTBT at that time was in keeping with the strategic requirements of the country then. Although it had not formally declared so in public, India was on track at that time to build a nuclear

8. https://www.ctbto.org/nuclear-testing/testing-times/
9. Robert S. Norris and William M. Arkin, "Known Nuclear Tests Worldwide, 1945-98", NRDC: *Nuclear Notebook*, November/December 1998. Vol. 54, No. 06, pp. 65-67. © 1998 *The Bulletin of the Atomic Scientists*.
10. Fact Sheets & Briefs, Arms Control Association. This chart includes tests done by India, Pakistan and North Korea. https://www.armscontrol.org/factsheets/nucleartesttally

arsenal of its own. Indeed, it had conducted the Pokhran II tests within a few years after pulling out of the CTBT negotiations.

But, this author personally believes that the time has now come for us to sign the CTBT.[11] We lose nothing, and we can claim our place as a leader and not just a follower in the community of nuclear nations, and that too without flexing any military or economic muscle. In any case, prior to signing the Indo-US Treaty, our Foreign Minister of the time, Mr Pranab Mukherjee had already announced a voluntary moratorium on nuclear tests[12], which helped clear the way for the Nuclear Suppliers Group (NSG) to lift the sanctions against India . It was well understood by all parties to the nuclear deal that any attempt on our part to conduct another nuclear test would not only threaten to nullify the nuclear deal but also lose us the hard won acceptability as a responsible nuclear power that we now enjoy.

More importantly, our sensible nuclear doctrine with its emphasis on a minimum nuclear deterrent does not require us to keep testing newer and more powerful weapons. The weapons we already possess are quite adequate for inflicting "unacceptable damage" on Pakistan or China. India does not need to build a thermonuclear weapon just because China has done so. This author has made these arguments several times in the public fora.[13] Therefore, apart from losing international acceptance and the nuclear deal, we also do not have a strategic need to test again under any foreseeable circumstances. So, why not join the CTBT and put the non-signatories and the non-ratifiers to shame?

## Confidence-Building Measures (CBMs)

Unlike arms control issues, for which the above sections have presented a detailed analysis of their strategic and technical underpinnings,

11. R. Rajaraman, "It Is Time India Signs the Nuclear Test Ban Treaty", *The Wire* (India), January 6, 2017. https://thewire.in/94539/india-nuclear-test-ban-treaty/
12. http://articles.economictimes.indiatimes.com/2008-09- 05/news/28441988_1_nsg-countries-waiver-moratorium-on-nuclear-testing
13. For a selected sample, see R.Rajaraman, "India-US Deal and the Nuclear Ceiling" *The Hindu*, Edit page, September 10, 2005; R. Rajaraman, "Cap the Nuclear Arsenal Now", *The Hindu*, Edit page, January 25, 2005.

Confidence- Building Measures (CBM) are relatively simpler to state and understand. Appreciating them requires only common sense. But this is not to underestimate their importance. It must be emphasised that CBMs contribute greatly to reducing nuclear dangers, as much as the various arms control measures.

In our subcontinent, India and Pakistan had set up some basic CBMs a long time ago. These include:

(i) Non-attack on each other's nuclear facilities.

(ii) Hotline between the Directors General of Military Operations (DGMOs) of the two countries. With possible skirmishes between the two countries, especially across the border, such a hotline will help prevent skirmishes from escalating into larger confrontations.

(iii) Memoranda of Understanding (MoUs) for advance notification of missile launches.

These specific measures are more or less extensions of conventional CBMs on nuclear facilities and nuclear missiles. In addition, there are more sophisticated CBMs specific to nuclear dynamics and strategy. The most prominent measure in that category is our declaration of No First Use (NFU) as one of the main pillars of our nuclear doctrine. Although Pakistani analysts like to tell us at international meetings that India's NFU has no credibility and does not affect Pakistan's level of preparedness, it must surely give them some measure of security. Of course, CBMs have to be mutual, but Pakistan categorically refuses to consider NFU.

In addition, Track II discussions between experts from the two nations can be a big CBM. They not only set up some personal contact with adversaries but also provide an opportunity to clarify matters. I recall discussing with Pakistani counterparts the possibility of Pakistan building a short range missile (erroneously called a "tactical weapon" by many analysts) before the Nasr was officially announced.

Apart from Track II meetings, continuous dialogue and mutual visits between officials of both sides dealing with strategic matters would also be a great help.

# 14. The Current State of Play in Nuclear Disarmament

*Swaran Singh and Reshmi Kazi*

Nuclear weapons comprise the most terrifying technology humans have ever discovered. In fact, in this human sojourn to continuously seek control over greater firepower, nuclear weapons have presented mankind with a historic dilemma where their 'non-use' (read deterrence) remains the only sensible use though nations have also been preparing for nuclear war-fighting as a part of building credible deterrence. Disarmament (or non-proliferation and arms control, short of disarmament) have been the other end of the deterrence stability spectrum. Nuclear disarmament indeed presents the only choice to avoid the prospects of intended or inadvertent extinction of life, and experts, activists and policy practitioners have long been campaigning for the elimination of such dangerous technologies. Accordingly, pursuing global disarmament was one avowed objective of the Treaty on Non-Proliferation of Nuclear Weapons (NPT) of 1971. But, the current state of play of nuclear disarmament remains a complex web of two steps forward and one step backward. Indeed, the recent history of nuclear disarmament has been marked by some novel and positive transformations in its structures and processes though the objective yet remains elusive.

The year 2022 began with a newfound optimism amongst disarmament proponents. In a rare gesture, all the five permanent members of the UN Security Council—that are also the five Nuclear Weapon States (NWS) so recognised by the NPT—issued, for the first time, a joint statement underlining "that a nuclear war cannot be won and must never be fought."[1] Similar words had also been stated in the bilateral joint statement of Moscow and Washington; first, from the Reagan-Gorbachev 1985 Summit and, more recently, from the Biden-Putin one in June 2021 in Geneva. The pertinent question, however, is whether these will continue to be lip-service aimed at creating atmospherics for resolving momentary geopolitical tensions or can they promise an inflection point for nuclear disarmament?

It is in this backdrop that this chapter revisits the current state of nuclear disarmament. While surveying the overall trends amongst the major stakeholders, it especially highlights two salient transformations. First, the lack of adequate commitment amongst the P5 to implement the NPT mandated nuclear disarmament which has eroded their credibility amongst the non-NWS (NNWS). This has motivated many NNWS to advocate nuclear disarmament at various multilateral fora beyond P5-led structures and initiatives. Second, it examines several civil society and other non-state actors like the Pugwash Movement, International Campaign to Abolish Nuclear Weapons (ICAN), International Council of the Red Cross, etc that have reemerged as leading players piloting nuclear disarmament campaigns and introducing new tools, techniques and outcomes.

## Highs and Lows of Nuclear Disarmament

After the world witnessed catastrophic destructive power with the first use of two atomic weapons at Hiroshima and Nagasaki in 1945, the UN General Assembly (UNGA) debated and adopted resolutions

1. The White House, "Joint Statement of the Leaders of the Five Nuclear-Weapon States on Preventing Nuclear War and Avoiding Arms Races", January 3, 2022. https://www.whitehouse.gov/briefing-room/statements-releases/2022/01/03/p5-statement-on-preventing-nuclear-war-and-avoiding-arms-races/. Accessed on January 4, 2022; Lewis Dunn and William Potter, "Time to Review the Reagan-Gorbachev Principle", *Arms Control Today*, March 2020. https://www.armscontrol.org/act/2020-03/features/time-renew-reagan-gorbachev-principle. Accessed on January 9, 2022. International Court of Justice, "Legality of the Threat or Use of Nuclear Weapons", July 8, 1996. https://www.icj-cij.org/en/case/95. Accessed on January 4, 2022.

for general and complete disarmament. Nevertheless, the arms race, especially between the United States and Soviet Union, was to increase their arsenals to over 70,000 warheads by the mid-1980s. The 1970s and 1980s were decades that saw major peace and nuclear freeze movements across Europe as also initiatives like the Action Plans presented to the UN General Assembly by India's Prime Minister Rajiv Gandhi. But these remained of little avail and were generally sidelined.

However, these decades also saw several treaties between the two superpowers, bringing their nuclear warheads to the current low of 13,080. At the same time, the sheer economic burden of the arms race triggered the beginning of the end of the Soviet Union. Meanwhile, proliferation was contained. In 1963, the Kennedy Administration had estimated that world would have as many as 25 NWS by the 1970s.[2] But 75 years after the discovery of nuclear weapons, the P5, along with their NPT-led non-proliferation regime, have ensured that only four—India, Israel, North Korea and Pakistan—were able to cross the nuclear rubicon. These four Nuclear Armed States (NAS) have relatively small stockpiles and are not recognised as NWS under the NPT. Then there are others that were either denied or disarmed using coercion and incentives, and yet others like Iran that continue to be seen as aspirants, and, therefore, are targets of the P5.

The onus for nuclear disarmament has, therefore, remained with the P5, or with the US and Soviet Union/Russia as these two even today hold more than 90 percent of the world's total nuclear stockpile. Their weapons are also the most diversely deployed around the planet as also the most advanced compared to other NWS or NAS. Moreover, while reducing their nuclear stockpiles, they have continued to modernise these weapons. As a result, even in the face of a strong norm of nuclear taboo and a fairly effective non-proliferation regime, the argument of the nuclear weapon being the ultimate currency of power continues to thrive. Moreover, given their destructive power, even their currently limited stockpiles are far too big; more than enough to destroy our world

2. Claudia Kissling, *Civil Society and Nuclear Non-Proliferation: How do States Respond?*, (New York: Routledge, 2016), p. 35.

several times over.[3] Indeed, their very existence, involving their upkeep, safety and modernisation, continues to divert crucial human, technical and financial resources, and erode the chances for building peace and development.

Undeniably, at least as a ritual, most inter-governmental meetings see the NWS and NAS express their commitment for eliminating nuclear weapons. The aims of nuclear disarmament are often propagated as the key priority by most NWS/NAS. But there remains wide disparity in their profession and practice; their nuclear disarmament visions remain undergirded by their nuclear deterrence strategies. The most apt example of this disconnect remains the failure of nearly half of the NPTs five-yearly Review Conferences (RevCons) to even produce a consensus document; and now the 10th RevCon of the NPT which was originally scheduled for March 2020, has been postponed repeatedly.[4] Indeed, the Chemical Weapons Convention of 1993 was the last major disarmament treaty negotiated by state actors and the initiative has since shifted to civil society organisations.[5]

In the face of the P5 being unable to sustain their disarmament negotiations or coopt the NAS in their efforts, multilateral nuclear disarmament has seen various middle powers from amongst the NNWS, non-governmental actors, retired officials and politician, and other iconic thought-leaders push disarmament through the United Nations General Assembly. Inspired by the success of the Anti-Personnel Mine Ban Convention (1997) and the Convention on Cluster Munitions (2008),

3. Antonio Guterres, "Our Main Hope to Reverse Course From Nuclear Cataclysm Is the Non-Proliferation Treaty", *The Wire*, January 3, 2022. https://thewire.in/world/nuclear-non-proliferation-treaty-negotiations-antonio-guterres-op-ed. Accessed on January 9, 2022.
4. Wilfred Wan, "Why the 2015 NPT Review Conference Fell Apart", UN University, Centre for Policy Research, May 28, 2015. https://cpr.unu.edu/publications/articles/why-the-2015-npt-review-conference-fell-apart.html. Accessed on December 10, 2021.
5. Swaran Singh, "P5 Should See Civil Society as Key to Eliminating Nukes", *Asia Times* (Hong Kong), January 7, 2022. https://asiatimes.com/2022/01/p5-should-see-civil-society-as-key-to-eliminating-nukes/. Accessed on January 7, 2022.

several International Non-Governmental Organisations (INGOs) and their innovative new approaches have emerged as a force to reckon with.

The P5 have seen a slide in their own initiatives, making space for other actors to take the lead. The year 1996 saw the African Nuclear Weapon-Free Zone (NWFZ) Treaty (Treaty of Pelindaba) being opened for signatures and ratifications. This was an initiative of the African Union and not the P5; so much so that of the P5 took long as 15 years to sign and ratify it, with Russia being the last to ratify it in 2011. In 1996, the International Court of Justice pronounced its advisory opinion on the legality of the threat or use of nuclear weapons in any circumstance being permitted under international law.[6] While the court remained divided on the legality of the threat use of nuclear weapons, it unanimously announced that "there exists an obligation to pursue in good faith and bring to a conclusion negotiations leading to nuclear disarmament in all its aspects under strict and effective international control." In April 1997, a consortium of lawyers, scientists, physicians, former diplomats and disarmament specialists and activists, drafted a concrete and tangible model Nuclear Weapons Convention (NWC) that marked an inflection point in nuclear disarmament trajectories. The NWC was modelled on the 1993 Chemical Weapons Convention and the 1972 Biological Weapons Convention stipulating to prohibit all the signatory states from the development, testing, production, transfer, use and threat of use of nuclear weapons. This first draft had come a bit too early and it needed several more years of sustained civil society advocacy to finally become a reality as the Treaty for the Prohibition of Nuclear Weapons (TPNW) that came into effect in January 2021.

## Nuclear Disarmament: A Reality Check

Despite some of these aforesaid notable achievements, the focus of the two superpowers has remained on propagating disarmament

6. John Burroughs, "Looking Back: The 1996 Advisory Opinion of the International Court of Justice", *Arms Control Today*, July/August 2016. https://www.armscontrol.org/ACT/2016_07/Features/Looking-Back-The-1996-Advisory-Opinion-of-the-International-Court-of-Justice. Accessed on December 15, 2022.

from the prism of national security. Their efforts at denying nuclear technologies to others while constantly modernising their own seriously undermine nuclear disarmament. The 1996 CTBT was the last treaty to be debated at the Conference on Disarmament (CD) in Geneva, and even this treaty was put in the deep freezer with the US Senate rejecting it in October 1999.[7] This was followed by the failure of the 2005 NPT Review Conference to reach a final document.[8] If anything, this period saw the International Atomic Energy Agency (IAEA) further tightening control over the NNWS, all in the name of "non-compliance cases" that saw the introduction of intrusive Additional Protocols on the comprehensive safeguards "which expanded the verification responsibilities of the agency and each state-party", making the NNWS further sceptical of the nuclear disarmament intent of the NWS.[9] All this saw several NNWSs becoming shy of signing the Additional Protocols that called for stricter constraints on their civilian nuclear programmes. The unabated operation of illicit nuclear trafficking and the consequent revelation of the AQ Khan nuclear black market remains a quintessential example of how such activities have weakened the foundation of the non-proliferation regime and, thereby, impeded disarmament efforts.[10] The weakened NPT—the axis of the multilateral non-proliferation regime built around the P5's commitment for disarmament—crippled their ability to collectively set

7. Helen Dewar, "Senate Rejects Test Ban Treaty", *The Washington Post*, October 14, 1999. https://www.washingtonpost.com/wp-srv/politics/daily/oct99/senate14.htm. Accessed on December 12, 2021.
8. John Simpson and Jenny Nielsen, "The 2005 NPT Review Conference: Mission Impossible?", *The Nonproliferation Review*, Vol.12, Issue 2, 2005, https://www.tandfonline.com/doi/abs/10.1080/10736700500378901. Accessed on November 18, 2022.
9. Trevor Findlay, "Looking Back: The Additional Protocol", *Arms Control Today*. https://www.armscontrol.org/act/2007_11/Lookingback. Accessed on December 12, 2021.
10. IISS Strategic Dossier, *Nuclear Black Markets: Pakistan, A.Q.Khan and the Rise of Proliferation Networks: A Net Assessment* (London: The International Institute for Strategic Studies, 2007); United States Congress, US House of Representatives, Committee on International Relations, *The A Q Khan Network: Case Closed?*, 2018.

the tone and tenor of global nuclear disarmament narratives, leaving it to their disparate individual trajectories.

## The United States of America

The US clearly has been the world leader in both the range and modernisation of nuclear arsenals as also in leading nuclear disarmament narratives. The US' aspirations to preserve its global leadership, premised on maintaining robust nuclear deterrence, including its extended nuclear deterrence to protect its allies, has influenced its disarmament efforts. Hence, in spite of its pledge towards nuclear disarmament, even in the post-Cold War era, the US leadership remains anchored on nuclear deterrence that it credits for having preserved regional and international stability.

Amongst some of the recent examples of the US' pursuit for disarmament were those of President Barack Obama. Even as a senator, in his July 2008 speech at Purdue University, West Lafayette, Indiana, he had pronounced: "It's time to send a clear message to the world: America seeks a world with no nuclear weapons."[11] As president, he began his innings promising to strive towards the goal of total abolition of all nuclear weapons and was later conferred the Nobel Peace Prize. In his presidency's first big foreign policy speech to a gathering of tens of thousands at Prague in April 2009, Obama had outlined his vision where he saw the US as "the sole country to fire a nuclear weapon in anger" and, therefore, it "bore the moral responsibility for launching a new era of nuclear disarmament" as he pronounced his "commitment to seek the peace and security of a world without nuclear weapons."[12] Obama had even toyed with the idea of adopting 'No-First Use' (NFU) but it was stymied by the US' need to keep its allies in confidence. Obama's pursuit,

---

11. Alexander Mooney, "Obama Says Time to Rid World of Nuclear Weapons", CNN Politics, Election Centre 2008. https://edition.cnn.com/2008/POLITICS/07/16/obama.speech/. Accessed on November 30, 2021.
12. Ian Traynor, "Barack Obama Launches Doctrine for Nuclear-Free World", *The Guardian* (London), April 5, 2009. https://www.theguardian.com/world/2009/apr/05/nuclear-weapons-barack-obama. Accessed on January 5, 2022.

therefore, remained cautiously calibrated in the goal that till global nuclear disarmament is established, the US will continue to maintain a safe, secure, and reliable arsenal to deter any adversary, and assure the security of America's allies.

The central tenets of US nuclear disarmament, therefore, have remained embedded in the principles of nuclear deterrence and extended nuclear deterrence to its allies. Even the most recent nuclear policy and force posture of the Biden Administration, the *Interim National Security Strategic Guidance* of March 2021 says that it "will take steps to reduce the role of nuclear weapons in national security strategy, while ensuring strategic deterrence remains safe, secure and effective and that extended deterrence commitments to allies remain strong and credible."[13] Till Biden's *Nuclear Posture Review* gets published, the last one issued in 2018 by President Trump recommends maintenance and modernisation of a nuclear triad and its associated nuclear command and control and supporting infrastructure; keeping options open against China's expanding build-up, and quantitative improvements in its strategic nuclear forces; pursuing the development and acquisition of nuclear-armed sea-launched cruise missiles; and rejection of declaratory nuclear policies of no first use.[14] Meanwhile, the nuclear hawks have already begun to pressurise the Biden Administration against imposing curbs on the arsenal, including defence budget cuts on arsenal modernisation.[15]

---

13. Nobuyasu Abe, "The Biden Nuclear Posture Review and Its Implications for US Asia-Pacific Allies", *Journal for Peace and Nuclear Disarmament*, Vol. 4, Issue 2, 2021, p. 235.
14. Office of the Secretary of Defence, US Department of Defence, *Nuclear Posture Review*, February 2018. https://media.defence.gov/2018/Feb/02/2001872886/-1/-1/1/2018-NUCLEAR-POSTURE-REVIEW-FINAL-REPORT.PDF. Accessed on January 11, 2022.
15. Julian Borger, "Nuclear Arms Hawks Give Bureaucratic Mauling to Biden Vow to Curb Arsenal", *The Guardian* (London), November 3, 2021. https://www.theguardian.com/world/2021/nov/03/nuclear-arms-joe-biden-pentagon-hawks. Accessed on November 25, 2021.

## Russian Federation

In 1993, Russia inherited over 32,000 nuclear warheads which it has now reduced, under the New Strategic Arms Reduction Treaty (START) Treaty, to 1,550.[16] However, all US-Russian arms control and disarmament initiatives have remained premised on creating nothing more than "mutual predictability" and "to maintain parity" while simultaneously "modernising their nuclear forces."[17] Also, while some US Administrations have sometimes shown an inclination for adopting an NFU doctrine, Russia, from the very beginning in 1993 had dropped this pledge that was first pronounced by Leonid Brezhnev in 1982.[18] The Soviet Union, of course, had been a torchbearer of both the NFU as also the nuclear test ban which are often described as the first essential steps towards nuclear disarmament.

Second, the rise of China has also contributed to Russia's relative decline as a world power. This has seen President Putin "trumpeting new and exotic systems" meant "to show that Russia is still a great power to be reckoned with."[19] Russia has also become deeply concerned about the

---

16. Marco De Andreis and Francesco Calogero, *The Soviet Nuclear Weapons Legacy* (Stockholm, SIPRI Research Report No. 10), p. 3.
17. Rose Gottemoeller, "Russia Is Updating Their Nuclear Weapons: What Does That Mean for the Rest of Us?", Carnegie Endowment for International Peace, January 29, 2020. https://carnegieendowment.org/2020/01/29/russia-is-updating-their-nuclear-weapons-what-does-that-mean-for-rest-of-us-pub-80895. Accessed on January 12, 2022.
18. P M Kamath, "The First and No First Use Options of Nuclear Weapons", *World Affairs: The Journal of International Issues*, Vol. 17, No. 4, October-December 2013, p. 27; Rajiv Nayan, "The Biden Administration and the Future of 'No First Use'", *IDSA Comment*, December 28, 2021. https://www.idsa.in/idsacomments/the-biden-administration-and-the-future-of-no-first-use-rajivnayan-28122. Accessed on January 2, 2022; Vladimir Isachenkov, "New Russian Policy Allows use of Atomic Weapons Against Non-Nuclear Strike", *Defence News*, June 2, 2020. https://www.defensenews.com/global/europe/2020/06/02/new-russian-policy-allows-use-of-atomic-weapons-against-non-nuclear-strike/. Accessed on June 2, 2020.
19. Ibid.; Swaran Singh, "Hypersonic Missiles: A New Rocket Race", *Asia Times* (Hong Kong), January 13, 2022. https://asiatimes.com/2022/01/hypersonic-missiles-a-new-rocket-race/. Accessed on January 14, 2022.

US' conventional superiority—so effectively showcased during the first Persian Gulf War of 1991, making Moscow step back from its NFU doctrine and pronounce that it may use nuclear arms to repulse a non-nuclear attack. Third, this period has also seen the North Atlantic Treaty Organisation's (NATO's) eastward enlargement, with former Soviet Republics like Estonia, Latvia, Lithuania and the former countries of the Eastern block like Bulgaria, Romania, Slovakia and Slovania joining NATO. Indulgence by NATO with Georgia and Ukraine is perceived as a direct security threat by Russia, making it reinforce its nuclear deterrence.

So, Russia, like its predecessor, the Soviet Union, has continued to see nuclear weapons as guarantors of peace and sovereignty and symbols of great power status, and the primary purpose of its nuclear weapons remains strategic deterrence. No doubt, like the US, the end of the Cold War had seen Russia engaging in debates on reducing the salience of nuclear weapons. But, on nuclear disarmament, while Russia officially advocates the total abolition of nuclear weapons, it perceives global nuclear disarmament as an extremely complex aspiration. Russia has extended support to the CTBT and underscored nuclear reductions. But, any overt discussion on the prospects for a world free of nuclear weapons remains rare amongst its strategic community. Given its complex relationship with the US, nuclear weapons are expected to remain a predominant feature of Moscow's narratives.

### France

France once had the world's third largest nuclear stockpiles after the US and Soviet Union but has since substantially reduced its weapons—from 540 to around 290—including eliminating all land-based Intercontinental Ballistic Missiles (ICBMs). Nevertheless, given its bitter experiences during World Wars I and II, both the history of its *force de frappe*, in general, and its commitment to nuclear deterrence, in particular, have remained critical to safeguarding the vital interests and pride of France.[20] Indeed, the French nuclear forces have not been part of NATO's integrated military command structure and retain their independent decision-making. Also, other than the US, it is the only NATO power to possess its own nuclear-powered 42,000-ton aircraft carrier aptly named *Charles de Gaulle*.[21]

The French White Paper on Defence and National Security: NATO, Nuclear Weapons and Space of 2008, down lays the French strategy in a 15-year perspective "to reinvent its defence policy from one 'resting primarily on nuclear deterrence' to one more involved with military operations" and also "proposing a nuclear disarmament action plan, France will still rely on its nuclear deterrent."[22] So France has had reservations about an NFU, and its nuclear disarmament remains premised on nuclear deterrence that entails ensuring the integrity of its territory, the safety of its people and its sovereignty by threatening to deliver unacceptable damage in retaliation to an attack, "wherever if may come from, and in whatever shape and form."[23]

As regards its nuclear disarmament track record, other than the recent influence of the Greens, "it is worth noting that… there has never been, in France, any significant movement against nuclear weapons."[24] In the absence of strong domestic pro-disarmament constituencies, France has remained a conservative participant in nuclear disarmament

20. Nicolas Badalassi and Frederic Gloriant, "Introduction", in Nicolas Badalassi and Frederic Gloriant, eds., *France, Germany and Nuclear Deterrence: Quarrels and Convergences during the Cold War and Beyond* (New York: Berghahn Books, 2022), p. 1; Wolf Mendl, "French Attitude on Disarmament: Disarmament June 1967", *Survival* (IISS, London), Vol. 9, Issue 12, 1967, p. 393.
21. Kyle Mizokami, "Is the French Military Ready to Fight Any War?", *The National Interest*, January 8, 2022. https://nationalinterest.org/blog/reboot/french-military-ready-fight-any-war-199104. Accessed on January 9, 2022.
22. The French White Paper on Defence and National Security: NATO, Nuclear Weapons and Space, *European Security Review*, No. 40, September 2008, pp. 2, 5.
23. Prakash Menon, "Nuclear Doctrines and Military Realities", in V. R. Raghavan, ed., *Global Nuclear Disarmament: Geopolitical Necessities* (New Delhi: Vij Books, 2012), p. 46.
24. Anne Stevens, *The Government and Politics of France* (London: The Macmillan Press, 1992), p. 270.

discourses. Not surprisingly, it joined the NPT only in 1992 and the Biological Weapons Convention in 1984. However, the recent evolution of the European Union's (EUs) Common Foreign and Security Policy has provided a new frame for the French disarmament narrative, making it a bit more forthcoming. The French presidency of the EU in 2008, for instance, saw the EU adopt an action plan for the UN General Assembly proposing strengthening of a series of disarmament initiatives.[25] With Brexit, France is today the only NWS in the EU and President Macron has lately urged Europe to "rethink disarmament" by looking at the "unique track record" of France, citing its irreversible dismantlement of all its land-based nuclear weapons.[26]

So, while France has become relatively open to engage in disarmament discourses and undertaken selective unilateral force reductions, like the two superpowers, its disarmament remains premised on modernising its sea-based ballistic missile submarine force as also its airborne missiles carried by nuclear-capable bombers, thus, keeping nuclear weapons central to its deterrence strategies.

### United Kingdom

Quite unlike France, the nuclear research, rationale and force posture of the UK have remained closely intertwined with those of the US. Their partnership was first institutionalised in the1958 US-UK Mutual Defence Agreement and in their NATO alliance. The UK's nuclear disarmament vision remains premised in the belief that nuclear weapons are an integral part of its defence doctrine and that these weapons have played a crucial role not just in safeguarding against uncertainties and risks but in obtaining it crucial influence on the US and in NATO

25. Camille Grand, "France and Disarmament From One Century to Another", diploweb.com La revue geopolitique, July 25, 2020. https://www.diploweb.com/France-and-disarmament-from-one.html. Accessed on January 2, 2022.
26. Shannon Bugos, "France Offers Nuclear Deterrence to All Europe", *Arms Control Today*, March 2020. https://www.armscontrol.org/act/2020-03/news/france-offers-nuclear-deterrent-europe. Accessed on January 2, 2022.

strategies.[27] Having suffered in the face of a superior adversary like Germany and later the Soviet Union, the UK's nuclear policy strongly prioritises national security imperatives. The collapse of the Soviet Union had seen UK gradually reducing its nuclear stockpiles from about 500 warheads to under 200 warheads.[28] In March 2021, however, the UK decided to raise the ceiling of its nuclear warheads by 44 per cent—over and above its existing level of 180 warheads announced as per its 2010 *Strategic Defence and Security Review*.[29]

At the same time, the UK remains an ardent advocate of the global nuclear disarmament campaign and is a strong supporter of urgent ratification of the CTBT, calling for enhanced transparency, and achievable confidence-building measures among states; and has shown commitment to develop reliable technical conditions for effective nuclear disarmament verification. It has been a keen participant at the CD, has signed and ratified several Nuclear Weapons Free Zones (NWFZs) agreements and participated in initiatives like President Obama's Nuclear Security Summits as also the 2015 Joint Comprehensive Plan of Action or the Iran nuclear deal. However, this has not prevented London from keeping its deterrence policy open-ended and enhancing its nuclear stockpile for its security and as a source of continued international stature and prestige.

### People's Republic of China

The People's Republic of China (henceforth China) presents another interesting case. Till the early 1990s, it was an outlier to much of the

27. Ian Davis, *The British Bomb and NATO: Six Decades of Contributing to NATO's Strategic Nuclear Deterrence* (Stockholm: Stockholm International Peace Research Institute, 2015); Martin A. Smith, "British Nuclear Weapons and NATO in the Cold War and Beyond", *Internatioanl Affairs* (RUSI, London), Vol. 87, No. 6, November 2011, pp. 1385-1399.
28. Hans M. Kristensen, "United Kingdom Nuclear Weapons 2021", *The Bulletin of the Atomic Scientists*, Vol. 77, Issue 3, 2021. https://www.tandfonline.com/doi/full/10.1080/00963402.2021.1912309. Accessed on January 6, 2022.
29. Kingston Reif and Shannon Bugos, "UK to Increase Cap on Nuclear Warhead Stockpile", *Arms Control Today*, April 2021. https://www.armscontrol.org/act/2021-04/news/uk-increase-cap-nuclear-warhead-stockpile. Accessed on January 2, 2022.

global nuclear disarmament initiatives but since then, it has emerged as an integral and influential participant and player in setting its trajectories.[30] Communist China has historically viewed nuclear weapons from purely defensive and retaliatory perspectives, aimed at preventing nuclear blackmail and for the protection of its territorial sovereignty.[31] This was evident from its enunciation of an NFU policy pronounced by Chairman Mao in 1964.[32] Explicit US nuclear threats against its territory during the Korean War goaded China into developing its nuclear deterrent.[33] Secondly, successive leaders in Beijing have consistently supported "non-discriminatory disarmament and minimum deterrence" and offered to "engage on arms control only when US and Russian leaders achieve deeper cuts in their much larger nuclear arsenals."[34] China's nuclear disarmament policy has continued to place the onus on the US and Russia, saying, "As the two countries with the largest and most advanced nuclear arsenals in the world, the US and Russia bear primary responsibility for nuclear disarmament, and should earnestly implement existing treaties and further drastically and substantially reduce their nuclear stockpile, so as to enable other nuclear states to join multilateral negotiations towards the goal of complete nuclear disarmament."[35]

---

30. Nicola Horsburgh, *China & Global Nuclear Order: From Estrangement to Active Engagement* (Oxford: Oxford University Press, 2015), p. 148.
31. Kartik Bommakanti and Suyash Desai, "China's Nuclear Ambiguity and its Implication for India", ORF Occasional Paper, Issue No. 309, April 2019, p. 13; Hu Xijin, "China's Nuclear Power must be Boosted no Matter Where China-US Ties Head", *Global Times* (Beijing), February 5, 2021. https://www.globaltimes.cn/page/202102/1215064.shtml. Accessed on January 3, 2022.
32. Christopher P. Twomey, "China's Nuclear Doctrine and Deterrence Concept", in James M. Smith and Paul J. Bolt, eds., *China's Strategic Arsenal: Worldview, Doctrine, and Systems*, (Washington DC: Georgetown University Press, 2021), p. 47.
33. Ibid.
34. Daryl G. Kimball, "Engage China on Arms Control? Yes, and Here's How", *Arms Control Today*, June 2021. https://www.armscontrol.org/act/2021-06/focus/engage-china-arms-control-yes-heres-how. Accessed on January 2, 2022.
35. Zeng Rong, "Letter: China Rejects Calls to Join Nuclear Disarmament Talks", *Financial Times* (London), August 25, 2021. https://www.ft.com/content/634220f9-0d34-4e85-9290-698b333852e2. Accessed on January 2, 2022.

What is new is that for the US, a rising China has gradually replaced Russia as a major competitor though China's nuclear stockpiles remain small and least threatening. But its changed equations with the US have made China increasingly concerned with the unmatched superiority of US conventional and nuclear capabilities and especially its state-of-the art precision-guided munitions, space weapons and the decision to develop and deploy newer missile defences that can neutralise Beijing's limited second-strike assets. This has triggered China's overall military modernisation. Hence, the same old sense of securitisation and preservation of vital interests has evolved newer connotations in China's nuclear posture, including its showing greater commitment to US-led disarmament initiatives. China, for instance, has signed the CTBT and extended support to the FMCT negotiations. And then, like rest of the P5, China has also not signed the Treaty on the Prohibition of Nuclear Weapons (TPNW).

## Nuclear Disarmament Beyond the P5

Simply put, the security imperatives of the P5 have impeded nuclear disarmament. Their competitive mutual equations and resultant penchant for newer military technologies have sustained the nuclear arms race and fed their scepticism about disarmament. As a result, nuclear weapons continue to be produced, tested, refined, modernised and stockpiled by the NWS and NAS, and aspired for by others. The NPT has frozen an asymmetry between the NWS and the NNWS but as a temporary arrangement expecting the P5 nations to provide the NNWS with regulated and monitored access to civilian nuclear technologies while negotiating global nuclear disarmament among themselves. Understandably, the Cold War dynamics seriously corroded the prospects of the P5's disarmament and instead invigorated the doctrine of nuclear deterrence as the key to maintaining armed peace. But even with the end of the Cold War, as the global balance of power has shifted from a rigid bipolar to a briefly unipolar, and then to a fluid, multipolar world, the P5 have failed to collectively prioritise disarmament. The last half century of the NPT has seen only two attempts by the P5 to collectively express faith in the desirability of nuclear disarmament: the first was the September 24, 2009, special session of the UN Security Council (UNSC) convened by President Obama and more recently, their January 3, 2022 joint statement issued from the respective capitals.[36]

The post-Cold War period, however, has witnessed several civil society actors and middle powers amongst the NNWS taking the lead. But even their alternative strategies for nuclear disarmament are not possible without the collective participation of all the NWS, NAS and NNWS alike. Indeed, many have urged the P5 to start a dialogue for reducing the salience of nuclear weapons and to achieve arsenal reductions. The NPT, the largest multilateral mechanism that mandates the P5 to take the lead in nuclear disarmament has, however, exposed the P5's underlying tensions, divisions, and distrust.

## Middle Powers' Multilateral Diplomacy

The recent decades have seen multilateral disarmament diplomacy moving beyond the P5. Other than multiple sporadic efforts, more formal multilateral disarmament diplomacy initiatives can be traced to October 18, 2002, when South Africa had, on behalf of the Non-Aligned Movement (NAM), proposed a draft resolution titled "Promotion of Multilateralism in the Area of Disarmament and Non-Proliferation" at the First Committee of the UN General Assembly.[37] This effort resulted the UN General Assembly (UNGA) adopting Resolution 57/63 reaffirming that multilateralism was the core principle in negotiations

36. United Nations, Office for Disarmament Affairs, "Security Council Calls for World Free of Nuclear Weapons During Historic Summit". https://www.un.org/disarmament/institutions/security-council/summit-level/. Accessed on January 2, 2022; The White House, "Joint Statement of the Leaders of the Five Nuclear-Weapon States on Preventing Nuclear War and Avoiding Arms Race", January 3, 2022. https://www.whitehouse.gov/briefing-room/statements-releases/2022/01/03/p5-statement-on-preventing-nuclear-war-and-avoiding-arms-races/. Accessed on January 8, 2022.

37. "Promotion of Multilateralism in the Area of Disarmament and Non-Proliferation." *UNODA*. https://www.un.org/disarmament/topics/multilateralism/. Accessed on December 30, 2021.

related to non-proliferation and disarmament.[38] Since then, several such resolutions have been introduced and adopted in the UNGA. In December 2012, for instance, the UNGA adopted Resolution 67/56 to establish an Open-Ended Working Group to develop proposals to take forward multilateral nuclear disarmament negotiations for the achievement and maintenance of a world without nuclear weapons.[39]

Multilateral diplomacy has historically been effective for creating conducive atmospherics for dialogue, for reducing the trust deficit by mitigating mutual threat perceptions. Second, multilateral disarmament diplomacy has also found resonance in the norms-driven European Union, especially in its Global Strategy on Foreign and Security Policy (2016)[40] and the European Union Strategy Against Proliferation of Weapons of Mass Destruction (2003).[41] Likewise, the NAM and African Union have been the other ardent supporters of the multilateral disarmament diplomacy. Civil society actors, especially INGOs too have become major players in guiding the tone and tenor of, and transforming, multilateral, disarmament diplomacy's efficacy. But this shift in disarmament has come with its own set of challenges and opportunities.

## Civil Society and Nuclear Disarmament

The Cuban missile crisis of 1962 had created the first post-World War II spectre of a nuclear confrontation and triggered an era of East-West

38. "Promotion of Multilateralism in the Area of Disarmament, *UNGA*, Resolution 57/63 UNGA, December 30, 2002. https://www.un.org/ga/search/view_doc.asp?symbol=A/RES/57/63. Accessed on December 30, 2021.
39. "Taking Forward Multilateral Nuclear Disarmament," *UNGA*, Resolution A/RES/67/56, January 4, 2013. https://undocs.org/en/A/RES/67/56. Negotiations, January 4, 2013, at https://www.un.org/ga/search/view_doc.asp?symbol=A/RES/67/56. Accessed on December 30, 2021.
40. Annegret Bendiek, "The Global Strategy for the EU's Foreign and Security Policy", German Institute for International and Security Affairs, August 2016. https://www.swp-berlin.org/publications/products/comments/2016C38_bdk.pdf, pp. 1-13. Accessed on December 30, 2021.
41. "European Union Strategy against Proliferation of Weapons of Mass Destruction," Council of the European Union, 2003, at register.consilium.europa.eu/doc/srv?l=EN&f=ST%2015708%202003%20INIT. Accessed on December 30, 2021.

detente, resulting in multiple arms control treaties between the US and USSR. This period, however, also saw the rise of popular consciousness and people-driven nuclear freeze and peace movements producing mega public protests across European cities.[42] These protests presented the common people's aspirations, often facilitated by civil society organisations like the Quakers, Mennonites, or the Pugwash Movement to raise public awareness about Weapons of Mass Destruction (WMDs). But while these initiatives did influence nuclear narratives, they did not aim at making any policy interventions. Negotiating nuclear disarmament was seen as the remit and responsibility of sovereign nations. These campaigns, however, created constituencies for civil society's future policy interventions, making them potentially major stakeholders in nuclear disarmament. Continued negation and even defiance by the P5 of this civil society work was to make them more prominent. Amongst others, their alternative approaches were to soon achieve two major breakthroughs: the 1997 Anti-Personnel Mine Ban Convention (also known as the Ottawa Treaty) and the 2008 Convention on Cluster Munitions. This outcome later inspired civil society organisations in their work for a historic on the Treaty for Prohibition of Nuclear Weapons (TPNW) that, till December 2021, had been signed by 59, and ratified by 56 nations. But just like in the case of the aforementioned treaties of 1997 and 2008, none of the P5 has so far signed or ratified the TPNW.

### *(i) International Physicians for the Prevention of Nuclear War*

In the 1980s, US and Soviet doctors came together to create the International Physicians for the Prevention of Nuclear War (IPPNW) to raise public awareness about the devastating medical consequences of nuclear weapons. Presently, the IPPNW is a non-governmental federation of national medical groups in over 60 countries, representing doctors, medical students, frontline health workers, and like-minded

42. Henry Richard Maar III, *Freeze!: The Grassroots Movements to Halt the Arms Race and End the Cold War* (New York: Cornell University Press, 2022), p. 81.

people who share the common goal of creating a secure nuclear weapon-free world. It was conferred the 1985 Nobel Peace Prize in recognition of its achievements.[43] To cite the recent example of its advocacy, these health professionals in Australia, the UK and the US have expressed their concerns at Australia's proposed acquisition of nuclear-powered submarines with the UK's and US' assistance, stating that the plan will jeopardise global health and security.[44] In a joint statement, the IPPNW along with its affiliates—Medical Association for Prevention of War (Australia), Medact (UK) and Physicians for Social Responsibility (USA)—expressed their serious concern about an arms race and the health risks arising from the P5 countries' nuclear assistance to Australia.[45]

### *(ii) International Campaign to Abolish Nuclear Weapons*

Another turning point for the civil society-driven nuclear disarmament happened in 2005 when the IPPNW and Medical Association for the Prevention of War (MAPW) came together to advocate a "lateral thinking and a new approach to nuclear disarmament."[46] In mid-2006, at the IPPNW World Congress in Helsinki, council members unanimously recommended the establishment of the International Campaign to Abolish Nuclear Weapons (ICAN) to strive for the

---

43. Marion Birch and Leo van Bergen, "Towards Abolition: A Special Issue from the International Physicians for the Prevention of Nuclear War", *Medicine, Conflict and Survival*, Vol. 34, Issue 4, p. 237.
44. "Proposed US/UK Nuclear-Powered Submarines for Australia: Jeopardising Health and Fuelling an Arms Race," *IPPNW*, September 21, 2021. https://www.ippnw.org/wp-content/uploads/2021/09/Summary-of-joint-statement-AUKUS-21.9.21.docx.pdf, p. 1. Accessed on December 30, 2021.
45. "Proposed US/UK Nuclear-Powered Submarines for Australia Jeopardise Health While Escalating an Arms Race no one can Win," *IPPNW*, September 21, 2021. https://www.ippnw.org/wp-content/uploads/2021/09/IPPNW-statement-Nuclear-Subs-letterhead-21.9.21.pdf, pp.1-4. Accessed on December 30, 2021.
46. Dimitry Hawkins, Dave Sweeney and Tilman Ruff, "ICAN's Origins: From Little Things, Big Things Grow…", ICAN, October 2019. https://www.icanw.org/ican_origins. Accessed on January 9, 2022.

eradication of nuclear weapons.[47] The ICAN has since been an umbrella civil society organisation for nuclear-disarmament groups. The ICAN played a substantial role in the creation of the manuscript titled "Securing our Survival (SOS): The Case for a Nuclear Weapons Convention" submitted to the UN General Assembly in 2007. It engages and coordinates with a diverse range of groups, and works alongside the Red Cross to reshape the debate on nuclear weapons and rally momentum towards their elimination. ICAN, for instance, was the civil society coordinator for each of the three conferences in Norway, Mexico and Austria during 2013-14, focussing on the humanitarian impacts of nuclear detonations. More recently, it was a major force behind the conclusion of the UN Treaty on the Prohibition of Nuclear Weapons (TPNW) and it continues to work for its universal acceptance, including by the P5 countries.

### *(iii) International Committee of the Red Cross*

The International Committee of the Red Cross (ICRC)—three times winner of the Nobel Prize—whose origins go back to the late 19th century, has contributed to a series of historic treaties, from the Geneva Conventions of 1864, the Geneva Protocol of 1925, all the way to the Landmine Ban Treaty of 2008, Cluster Munitions Treaty of 2008 and Treaty on the Prohibition of Nuclear Weapons of 2021.[48] In April 2010, ICRC President Jakob Kellenberger affirmed that the movement "firmly believes that the debate about nuclear weapons must be conducted not only on the basis of military doctrines and power politics."[49] The ICRC has since worked hard to reshape nuclear risks as a humanitarian issue that demands urgent attention from the international community. It calls on states to pledge not to ever use nuclear weapons and to commit

47. Ray Acheson, *Banning the Bomb, Smashing the Patriarchy* (London: Rowman & Littlefield, 2021), p. 56.
48. Ritu Mathur, *Red Cross Interventions in Weapons Control* (New York: Lexington Books, 2017), p. 138.
49. Jakob Kellenberger, "Bringing the Era of Nuclear Weapons to an End," *ICRC*, April 20, 2010. https://www.icrc.org/en/doc/resources/documents/statement/nuclear-weapons-statement-200410.htm

themselves to prohibit their use and eliminate them through a binding agreement. The ICRC too was critical to the conclusion of the TPNW

### *(iv) International Partnership for Nuclear Disarmament Verification and Others*

The International Partnership for Nuclear Disarmament Verification (IPNDV) is another ongoing initiative that includes more than 25 NWS and NNWS.[50] In the face of increasing question marks on the performance of the International Atomic Energy Agency (IAEA), the IPNDV was created in 2014 to engage a diverse group of states to develop innovative nuclear disarmament monitoring and verification solutions. It seeks to develop international capacity and expertise on nuclear disarmament monitoring and verification and highlights the importance of verification in future reductions of nuclear weapons.

### *(v) The Hibakushas*

The survivors of the atomic bombing of Hiroshima and Nagasaki comprise another civil society force who have regularly been submitting the "Appeal of the Hibakusha" to the UN General Assembly, supported by tens of thousands of signatures.[51] There have also been individual actors, inter-governmental commissions or smaller groups of former officials periodically becoming active in making appeals for nuclear disarmament.[52] The 2009 Report of the International Commission on Nuclear Non-Proliferation and Disarmament, for instance,

---

50. For details, see https://www.ipndv.org/events/2021-end-of-year-meeting-phase-iii-interim-review/
51. United Nations, Office of Disarmament Affairs, "Appeal of the Hibakusha: More than 10.5 Million Signatures Supporting Call for the Elimination of Nuclear Weapons", October 18, 2019. https://www.un.org/disarmament/update/the-handover-of-the-appeal-of-the-hibakusha-more-than-105-million-signatures-supporting-call-for-the-elimination-of-nuclear-weapons/. Accessed on January 2, 2022.
52. Sverre Lodgaard, "Pathways to Nuclear Disarmament: Timeline Challenges", in V.R. Raghavan, ed., *Global Nuclear Disarmament: Geopolitical Necessities* (New Delhi: Vij Books, 2012), p. 1.

had propagated universalising NFU as a first step towards nuclear disarmament.[53]

### *(vi) New Action Coalition*

The New Action Coalition (NAC) was formed in 1998 to bridge the North-South divide. The initiative was led by Brazil, Egypt, Ireland, Mexico, New Zealand, South Africa and Sweden to work for a nuclear weapon-free world. Since 2000, it has been regularly presenting proposals at the NPT Review Conferences and also to the UN First Committee on Disarmament. All these nations had joined the Humanitarian Initiative in 2013 that worked to explore the humanitarian impact of nuclear weapons.

There are scores of civil society organisations that have since been formed, and are active in taking forward the new non-state narratives and initiatives for nuclear disarmament. Indeed, in the face of the expanding influence of globalisation, the increased stress on democratisation of domestic and global governance has opened new avenues for creating external as well internal pressure points to seek change in the NWS' nuclear perspectives. Experts now talk of the next stage being the "institutional democratisation of the NWS—involving the state, civil society and the public, for example—and how these may be developed to support disarmament action."[54]

## Treaty on the Prohibition of Nuclear Weapons (TPNW)

The TPNW[55] is the most recent disarmament milestone that came into effect from January 2021. It marks the most apt example of disarmament driven by civil society advocacy and their sustained campaigns through the UN General Assembly. The story of the TPNW goes back to

53. Swaran Singh, "No First Use as Strategic Doctrine", in Raghavan, ed., n. 52, p. 68.
54. Tim Street, *The Politics of Nuclear Disarmament: Obstacles to, and Opportunities for, Eliminating Nuclear Weapons* (New York: Routledge, 2021), p. 14.
55. For the Text of the TNPW see https://treaties.un.org/doc/Treaties/2017/07/20170707%2003-42%20PM/Ch_XXVI_9.pdf. Accessed on January 10, 2022.

the 2010 NPT Review Conference Final Document expressing "deep concern at the continued risk for humanity represented by the possibility of these weapons."[56] This coincided with humanitarian law emerging as the new frame for nuclear disarmament.[57] When in March 2017, the draft resolution for the TPNW was presented at the UN General Assembly, a large number of states voted in its favour.

The TPNW represents a manifestation of decades of debates, efforts, and campaigns of non-state actors, redefining both the process as also the outcome of disarmament diplomacy. It marks a shift from the P5-led treaties which have been marred by bias and discrimination. For example, compared to the NPT, the TPNW presents a new approach to delegitimise nuclear weapons. Its provisions call for a categorical prohibition of development, possession, deployment and use or threat of use of nuclear weapons or other nuclear explosives (Article 1). In doing so, it not just reinforces the NPT mandate on nuclear disarmament under Article VI, it also seeks to reduce the salience of nuclear weapons and exhorts nations to step up action to reduce nuclear risks and promote disarmament.

The TPNW, in addition, also stipulates the obligation to assist the victims of the use and testing of nuclear weapons and to redress affected environments (Article 7). However, the P5 NWS and NAS have so far continued to oppose it, calling it "an unrealistic approach to disarmament that threatens to disrupt the NPT regime."[58] And, without the NWS and NAS joining it, the TPNW effectively means little. A cautious assessment

56. Final Document, 2010 Review Conference of the Parties to the Treaty on the Non-Proliferation of Nuclear Weapons, p. 12. https://www.nonproliferation.org/wp-content/uploads/2015/04/2010_fd_part_i.pdf. Accessed on January 5, 2022.
57. Tytti Erasto et al., *A Fresh Breeze for Nuclear Disarmament in Europe?: Making Best use of the Treaty on the Prohibition of Nuclear Weapons* (Berlin: The Greens/EFA in the European Parliament, June 2021), p. 13.
58. Jimie Kwong, "Rescuing a Fraying Nuclear Nonproliferation Regime", Carnegie Endowment for International Peace, January 13, 2022. https://carnegieendowment.org/2022/01/13/rescuing-fraying-nuclear-nonproliferation-regime-pub-86189. Accessed on January 13, 2022.

would underline the uphill challenge that civil society INGOs face in convincing the P5 to join the TPNW.

### Future of Nuclear Disarmament

The end of the Cold War had witnessed a certain 'peace dividend' rhetoric. But, then a gradual drift ensured that the threat of nuclear weapons did not disappear. States, however, must realise that nuclear disarmament can be ignored only at great peril for the future of humankind.

What is especially disconcerting is that while civil society INGOs have built strong networks and track records to promote nuclear disarmament, the NWS and NAS have continued with extensive and expensive modernisation programmes, increasing the precision and potency of their reduced stockpiles, and building newer delivery systems. The bulk of their arms control, as always, has remained a camouflage for phasing out older inventories. China, France and the UK are actually increasing their weapons stockpiles. Thus, while there is no doubt that the US and Russia have achieved substantial reductions in their nuclear stockpiles, this recent upsurge amongst the smaller NWS and NAS and the continued mutual distrust amongst all of them presents a formidable challenge. The current state of play in nuclear disarmament can at best be termed cautiously optimistic.

# 15. Prospects of Nuclear Disarmament after the Russia-Ukraine Conflict: Could New Opportunities Arise?

*Manpreet Sethi*

President Putin's nuclear signalling during the Russia-Ukraine conflict has drawn attention to nuclear weapons in new ways. With the end of the Cold War, it had been presumed that nuclear weapons would fade out as the rivalry between the superpowers dissipated. However, three decades since then, relations between the US and Russia are as strained as they were during most of the Cold War and nuclear weapons are as entrenched as ever in national security strategies. Meanwhile, the number of nuclear weapons possessing states has grown, multiplying the nuclear dyads that have adversarial relations today.

There is no doubt that nuclear risks have grown due to stressed inter-state relations, breakdown of the arms control architecture, absence of strategic stability mechanisms and the galloping pace of technological advances that intersect dangerously with nuclear deterrence. Alarmed at the situation and frustrated by the non-action towards nuclear disarmament by the Nuclear Weapon States (NWS), 122 nations lent their support to a Treaty on the Prohibition of Nuclear Weapons (TPNW) that was negotiated and passed at the UN General Assembly (UNGA) in July 2017.[1] It prohibits development, testing, production,

1. The United Nations convened a conference to negotiate a legally binding instrument to prohibit nuclear weapons, leading towards their total elimination. The Assembly encouraged all member states to participate in the conference, along with international organisations and civil society representatives. The conference took

manufacture, acquisition, transfer, possession, and stockpiling of nuclear weapons, as well as their use or threat of use. After having secured 50 ratifications, the treaty entered into force in January 2021. However, its membership includes only Non-Nuclear Weapon States (NNWS) for now.

Anticipating pressure on themselves for their lack of initiative on nuclear disarmament, particularly at the Review Conference (RevCon) of the Treaty on Non-Proliferation of Nuclear Weapons (NPT), the P5 offered a Joint Statement on January 3, 2022.[2] They collectively voiced their commitment to nuclear disarmament and reiterated the 1987 Reagan-Gorbachev formulation that 'a nuclear war cannot be won and must not be fought'. However, the optimism that the statement generated and the hope that it would be followed up with relevant actions has dissipated in the context of the Russia-Ukraine conflict, which is already three months old at the time of writing this chapter. Resolution looks distant and conversations about the possibility of the use of nuclear weapons by Russia are in the air.

Against this backdrop, this chapter examines how some of the recent developments would impact the prospects of nuclear disarmament. Would Russia's nuclear signalling and military coercion towards a non-nuclear weapon state push countries towards nuclear weapons and away from disarmament? What kind of trends are likely to emerge? The first part of the chapter examines some likely scenarios that may pan out in the context of the conflict. The second section explains why the contemporary nuclear reality is so undesirable and dangerous. Flowing from there, the final section undertakes a comparative analysis of suitable pathways to a Nuclear Weapons Free World (NWFW) which could be stable and sustainable.

---

place from March 27 to 31, and from June 15 to July 7, 2017, in New York. The Text of the treaty is available at https://documents-dds-ny.un.org/doc/UNDOC/GEN/N17/209/73/PDF/N1720973.pdf?OpenElement. Accessed on April 10, 2022.

2. For the text of statement see White House Press Office, "Joint Statement of the Leaders of the Five Nuclear Weapon States on Preventing Nuclear War and Avoiding Arms Races". Available at https://geneva.usmission.gov/2022/01/03/p5-statement/. Accessed on April 11, 2022.

## Nuclear Trends Likely to Emerge from Russia-Ukraine Conflict

Russian military action against a non-nuclear weapon state has led to popularising the view that countries with nuclear weapons can behave with impunity against others. This is likely to prompt the NNWS to re-evaluate their security imperatives, especially where they face hostile nuclear weapon possessors. It is not surprising that some voices in Japan[3] and South Korea[4] have begun to argue in favour of possessing their own nuclear weapons. Nuclear sharing under the North Atlantic Treaty Organisation (NATO), which allows allies to jointly discuss, plan and train for nuclear missions is likely to remain the preferred choice of countries in Europe. It may be recalled that in the last few years, there has been an inclination towards calling upon the USA to withdraw nuclear weapons from the territories of Belgium, Germany, Italy, the Netherlands and Turkey.[5] Such demands are likely to be muted in view of the recent Russian behaviour. More countries, in fact, may want to acquire nuclear weapons. Meanwhile, Belarus amended its Constitution after a referendum on February 27, 2022, to allow Russia to station its weapons on its soil.[6]

Therefore, the conflict and Russia's behaviour are likely to erode the sense of security of the NNWS, including their faith in negative security assurances where these have been grudgingly provided by the NWS. Concerns to this effect will be heard at the NPT RevCon now scheduled in August 2022 and the divide between the two categories of states could grow. The manner in which this is handled will influence the future

3. Rupert Wingfield Hayes, "Will Ukraine Invasion Push Japan to Go Nuclear?", *BBC News*, March 26, 2022. Available at https://www.bbc.com/news/world-asia-60857346. Accessed on March 29, 2022.
4. Choe Sang Hun, "In South Korea, Ukraine War Revives the Nuclear Question", *New York Times*, April 6, 2022.
5. For instance, see, Pia Fuhrohp, Ulrich Kuhn and Oliver Meier, "Creating an Opportunity to Withdraw Nuclear Weapons from Europe", *Arms Control Today*, October 2020.
6. Reuters, "Belarus Referendum Approves Proposal to Renounce Non-nuclear Status: Agencies". https://www.reuters.com/world/europe/launchpad-russias-assault-ukraine-belarus-holds-referendum-renounce-non-nuclear-2022-02-27/, February 28, 2022.

course of nuclear non-proliferation. As it stands, the issue of nuclear proliferation does not look as well settled as it did only some years ago. And, the direction this takes will have repercussions for the prospects of nuclear disarmament.

Given President Putin's propensity for taking actions that were considered unthinkable only some months ago, it is difficult to assert that the possibility of deliberate use of nuclear weapons by Russia is zero. No one had imagined that a full-scale land invasion, in violation of the sovereignty and territorial integrity of another nation, was possible in Europe in the 21st century. But Putin made it happen. Now, would he go as far as using nuclear weapons in a premeditated, rationally thought-out action?

Going purely by logic, the actual use of nuclear weapons looks problematic owing to the difficulty in finding the right target. If Russia chooses to use them in Ukraine, a country that Putin wants to annex, then it would amount to capturing a radioactive wasteland that will have to be looked after for years. Also, a little forgotten fact is that Ukraine has been under the nuclear protection of China since 2013. The Joint Statement of China and Ukraine on Further Deepening Strategic Partnership states, "China pledges to unconditionally not use or threaten to use nuclear weapons against Ukraine, which is a non-nuclear-weapon state", It further goes on the state that if Ukraine was to be "invaded by the use of nuclear weapons or in the event of a threat of such aggression, provide Ukraine with corresponding security guarantees."[7] So, Russian nuclear use against Ukraine will bring Russia into conflict with China which is its ally and closest friend in these times.

Meanwhile, if Russia uses nuclear weapons against another European nation that is a NATO ally, it would bring the wrath of nuclear NATO upon itself. Nuclear retaliation, then, would nullify the gains Russia may have sought with its first use. So, well thought out use of nuclear weapons to gain politico-military objectives appears quite

7. Joint Statement of China and Ukraine on Further Deepening Strategic Partnership. Available on http://www.gov.cn/jrzg/2013-12/05/content_2543057.htm. Accessed on April 16, 2022.

difficult and even dangerously foolish. However, the inadvertent use of the weapon due to miscalculation, misperception or accident cannot be dismissed. Given that tensions are high, and disinformation campaigns are intensifying the fog of war, the country's actions may make it stumble into nuclear use.

In case nuclear weapons do get used, deliberately or inadvertently, it could impact the future of nuclear disarmament in two ways—pushing it away or bringing it closer. The sentiment that is generated in favour of one or the other of these scenarios would be determined by the actual damage caused to the people, and the socio-economic and environmental consequences of nuclear use.

One possible scenario that could distance the world from the realisation of nuclear disarmament could be the use of a small, low yield nuclear weapon against a primarily military target that goes unanswered. While this would certainly ensure containment of the nuclear conflict and, thus, rescue the world from severe damage, such a response, nevertheless, would carry the risk of normalising the use of such weapons in future conflicts too. This would be a particularly scary scenario since 'conventionalising' the use of nuclear weapons would open a pandora's box on the idea of proportionate nuclear war-fighting. Once the norm of nuclear non-use is broken, the possibility of more states resorting to such use will raise the frightening risk of escalation in each conflict, with no guarantee that it would be contained in the case of each crisis. This would place the world in a rather precarious position.

Most other outcomes of the conflict, however, are likely to trigger a deep abhorrence for nuclear weapons and prompt nations towards their elimination. In case Russia's nuclear use provokes a nuclear retaliation from the West, the consequences could be far-reaching. Those who survive the exchange would want to move towards disarmament so as not to risk another such hair-raising episode.

A nuclear exchange would also demolish the longstanding, often made, claim that these weapons ensure stability and peace between nuclear-armed nations. Nuclear deterrence is believed to have averted a world war by "preventing a Cold War from turning into World War

III because they induced both Washington and Moscow to be more prudent… and led to the establishment of tools for crisis management to reduce the prospect of the outbreak of unintended warfare, either nuclear or conventional."[8] The Russia-Ukraine conflict, leading to a breakdown of nuclear deterrence, could undercut such positive spin around nuclear weapons. Nations will be forced to recognise the dangers of World War III created by the presence of such weapons and the heightened possibility of their use once the nuclear taboo has been broken.

Meanwhile, there could be a third scenario wherein the conflict may end without an actual use of nuclear weapons. If this happens, some rather heartening lessons may emerge in favour of nuclear disarmament since it would show up the non-utility of nuclear weapons for achieving any worthwhile politico-military objectives through premeditated use. It would demonstrate the difficulties in finding the right target for nuclear weapons, especially when two nuclear weapon states have assured second strike capabilities. Therefore, for all Russia's nuclear brinkmanship and bluster, which is an important dimension of nuclear deterrence, finding the appropriate military use for the weapon will not be that easy. This will strengthen the norm of non-use and reinforce the understanding that the nature of the weapon makes it suitable only for deterrence through retaliation.

Such a realisation could then be used to loosen the grip of nuclear weapons on national security strategies. But for this to happen, nations will also have to realise the dangers of the contemporary nuclear reality that we seem to be living in. Whichever way the ongoing conflict ends, the international community needs to comprehend the dangers inherent in today's nuclear reality. The following section holds a mirror to the kind of nuclearised world that we today inhabit.

## Contemporary Nuclear Reality: Can Dangerous be Desirable?

During the long years of the Cold War, bipolar nuclear confrontation was premised on mutual assured destruction. Both sides relied on

8. Richard Falk and David Krieger, *The Path to Zero: Dialogues on Nuclear Dangers* (Boulder and London: Paradigm Publishers, 2012), p. 26.

their vulnerability to each other to ensure deterrence stability. Even though the ideological confrontation between two heavily nuclear armed nations had its moments of nail-biting suspense, experiencing the dangers engendered a common desire to establish guardrails in the form of arms control mechanisms.

From then to now, nuclear weapons have moved into the arsenals of nine countries, leading to a very different situation. The first marked dissimilarity in the contemporary situation is the multiplicity of deterrence equations. Many nuclear dyads impinge on each other, thereby complicating stability dynamics. For instance, what China perceives as a necessary capability to ensure strategic stability with the USA, impinges on India's security, and India's response then impacts Pakistan. Obviously, a search for strategic stability amongst multiple overlapping dyads generates an unprecedented complexity.

Even more problematic is the fact that not all deterrence equations are imbued with the same sense of understanding of how nuclear deterrence works. Nuclear capability build-up is happening across nations. Meanwhile, strategic ambiguity is the hallmark of some doctrines and the concomitant lack of clarity tends to generate a hedging strategy in the adversary. Each assumes the worst of the other. Consequently, offensive and defensive capabilities are simultaneously being developed, acquired and enhanced. None of the dyads indicates an intention to undertake bilateral or multilateral arms control or confidence-building measures.

Meanwhile, in Asia, which has emerged as the "most nuclearised of all continents"[9], there is a rather complicated web of inter-state relations. Cohabiting the region are countries recognised as nuclear weapon states by the NPT, those that are nuclear armed but outside the NPT, those with the potential to go nuclear, those that could be tempted to acquire weapons themselves if that was to happen, as well as those that have been engaged in wilful nuclear proliferation.

9. Brad Roberts, "Asia's Major Powers and the Emerging Challenges to Nuclear Stability Among Them", Institute for Defence Analyses, Paper No. P-4423, February 2009, p. S-1.

Some other uncomfortable realities of the nuclear dyads in Asia are also evident. Firstly, most of these nuclear armed nations are geographically contiguous, which is in marked contrast to the US-USSR situation where the two were physically separated by an ocean. Secondly, most of the Asian nuclear armed states still have simmering territorial conflicts. Border skirmishes over undefined boundaries are routine, especially so in the case of India and Pakistan or India and China. The risk of escalation, therefore, is not negligible. Thirdly, espousing diverse nuclear doctrines, some of these countries believe in following a strategy of nuclear brinkmanship as a way of enhancing deterrence. This, however, is fraught with the risks of inadvertent nuclear escalation as a result of a miscalculation. For instance, both Pakistan and North Korea believe that they reap greater dividends from maintaining and even instigating strategic instability rather than supporting bilateral or multilateral measures that could bring a modicum of stability to their nuclear relationships. Fourthly, the region is also plagued by the scourge of terrorism and Pakistan, in fact, has demonstrated that a country with nuclear weapons can use them as a shield to foist acts of terrorism on another nuclear armed adversary. Political instability in nuclear weapon possessors that also harbour and support terrorist organisations raises the risk of nuclear terrorism for the entire international community.

As is evident then, the contemporary nuclear situation presents itself as a complex web of diverse strategic relationships with no common understanding of strategic stability, nor a uniform desire for addressing issues that are contributing to instability. The situation is certainly quite fraught with dangers of all kinds—deliberate nuclear war, inadvertent nuclear exchange, existential dangers from nuclear weapons, as well as the risk of nuclear terrorism. Much like the Cuban missile crisis 60 years ago, the most recent Russia-Ukraine conflict has once again shown the dangers of continuing to live with nuclear weapons. And, this time, the risks are even more pronounced owing to the multiple factors recounted above. The undesirability of the status quo is obvious, and so is the need to find ways of ridding the world of nuclear weapons.

## Possible Pathways to Nuclear Zero

In case the nuclear weapon states emerge from the Russia-Ukraine conflict with the understanding that the nuclear risks are untenable, there must be on offer a set of measures that can be envisioned to ensure a future world order that can provide for stability without nuclear weapons. The Russian behaviour has demolished the assumption that nuclear weapons can ensure stability. Rather, it has shown that such weapons can embolden military action. This should be unacceptable against the NNWS, and would be most dangerous against the NWS. Since wars cannot be outlawed, it should at least be imperative to eliminate weapons that might lead to unacceptable damage beyond the areas of conflict. The presence of nuclear weapons keeps mankind in a state of suspended fear of possible extinction every time a crisis arises between nuclear armed nations. As brought out by a strategic analyst writing about nuclear weapons, "The weapons we task with securing our nation are the same weapons that hold the potential to destroy other nations and to trigger an attack that will destroy our own."[10]

Therefore, it is necessary that international security be premised on a world without nuclear weapons. The routes to this destination can be many. From the current literature available on the subject, three different routes can be identified and these are examined in the following sub-sections.

### *Route 1: Via Reducing Numbers*

One route to an NWFW favours a progressive reduction in national nuclear arsenals. It is argued that as numbers go down, starting first with the USA and Russia that hold the bulk of the global nuclear stockpile, and eventually getting other smaller holders to join in, it would eventually lead to a state of zero nuclear weapons. Generally and intuitively speaking, reduction in nuclear numbers should be a welcome development. But for reductions to really count as a move towards zero, it needs to be seen whether they form a part of a larger plan, or whether

10. Falk and Krieger, n.1, p. 38

they merely amount to rationalising arsenals in keeping with altered threat perceptions?

Unless the reduction in numbers is accompanied by a simultaneous reduction in the salience of nuclear weapons through a revision of doctrines, it would be difficult to have stability at lower numbers. In fact, it could even be argued that mutual wariness could grow at lower numbers since countries may feel more vulnerable to a first strike.

In view of the above, it is unlikely that without a complementary reduction in the salience of nuclear weapons through the acceptance of doctrines that devalue or severely constrict the role of nuclear weapons, stability on the way to zero through a mere progressive reduction in numbers can be maintained.

### *Route 2: Via Wither-on-the-Vine*

A second approach to an NWFW suggests waiting it out for nuclear weapons to become obsolescent. Ward Wilson, a strategic analyst, explained this approach in an article entitled "The Pennyfarthing H-Bomb", published in the *World Today* in 2012. Targeting arguments made against the elimination of nuclear weapons because these can never be disinvented, Wilson argues that "technology does not go out of existence by being disinvented… [but when] better technology comes along, or it turns out it wasn't very smart technology after all." And, he finds that nuclear weapons have never been good military technology being as they are "too big, too clumsy, too messy and too outmoded to be any use to anyone."[11] As suggested earlier in this chapter, the Russia-Ukraine conflict may show up the military non-utility of nuclear weapons. If the episode ends without the use of these weapons, the arguments around their being morally reprehensible and legally unethical could be made to become stronger, making them more and more unusable. Eventually then, the weapon will wither-on-the-vine and become useless over a period of time.

---

11. Ward Wilson, "The Pennyfarthing H-Bomb", *The World Today,* October/November 2012, p. 26.

The problem with this route, however, is that while the international community waits for the weapons to 'wither', it would mean living with the serious dangers explained in the earlier section. And, these dangers are multiplying. Obviously then, other meaningful steps must be found to ensure stability as we get to zero and stay there. The following pathway may then serve as useful.

*Route 3: Via Leapfrogging Current Nuclear Beliefs*

Even as the nuclear warhead numbers reduce and the world waits for the weapons to become obsolescent, it will be worthwhile to adopt two guardrails. These can help manage the dangers while the weapons still exist and verifiable ways are found to destroy them. The two steps recommended are an acceptance by all NWS of a No First Use (NFU) agreement against the NWS and, a ban on the use or threat of use of nuclear weapons against the NNWS. The former will provide a sense of assurance and stability amongst the NWS, and the latter would provide reassurance to the NNWS. Most importantly, these steps will not seek to strip the NWS of their nuclear arsenals immediately. Given the kind of attachment that these states have shown towards their nuclear weapons, a global NFU and no-use treaties would allow them to retain the weapons for security, but it would become more and more notional over a period of time as the weapons fall into disuse. The benefits of the two approaches are briefly discussed below.

Credible first use of nuclear weapons requires a large nuclear arsenal, accurate and if possible Multiple Independently Targetable Reentry Vehicle (MIRV) missiles, an elaborate command and control structure, including a certain pre-delegation of authority to maintain a high level of readiness for use and an effective ballistic missile defence and its attendant infrastructure. Russia has taken this to a new level with the development of an unmanned nuclear-powered and nuclear-warhead-bearing torpedo such as the Poseidon.[12] The dangers of such moves, including of accidental use, are self-evident. Meanwhile, despite such an elaborate capability, a

12. "Poseidon: Unmanned Underwater Vehicle". https://www.military-today.com/navy/poseidon.htm. Accessed on April 1, 2022.

first user must still contend with the risk of retaliation when faced with another nuclear armed state that has a secure second-strike capability. In such a situation, first use may turn out to be suicidal.

On the other hand, the capability requirements of a state with an NFU are drastically reduced since it only has to promise assured retaliation to cause unacceptable damage. This can be well carried out with inaccurate missiles that can deliver even a few warheads over mega-cities. Neither do the command-and-control structures have to be elaborate enough to coordinate the logistics of a large first strike over many targets. Most of all, with a no first use doctrine, the decision-maker authorising nuclear use is freed of the psychological burden of having to make the momentous decision that would vaporise millions of people in minutes.

From both the military and political points of view then, NFU offers advantages, including those of reducing the risks of inadvertence. It allows even the first user to have a relatively relaxed posture since he is not under pressure of having to use his nuclear weapons early lest he loses them to preemption. In fact, a universal NFU would be even more relevant as nuclear weapons reduce. With small nuclear forces, the temptation to launch a disarming first strike would be high because of the 'use them or lose them' compulsions. But an NFU commitment would remove this temptation for both—self and adversary.

Adoption of NFU would be a crucial step towards the eventual elimination of nuclear weapons since it would involve an assurance from every country that it would not be the first to introduce nuclear weapons into a conflict. Since there would not be a first use, it would effectively mean no use of the nuclear weapon and, hence, a reduced dependence on the weapon in national security strategies over a period of time. This would reduce inclination towards horizontal proliferation by sending a strong signal of the diminishing utility of nuclear weapons; and it would lessen the drive towards vertical proliferation by reducing the need for the NWS to develop and deploy modernised nuclear arsenals. All of this would lower inter-state tensions, and contribute towards international security.

The second useful step to take would be to ban the use or threat of use of nuclear weapons. India has been a votary of this approach as encapsulated in the draft resolution entitled "Convention on the Prohibition of the Use of Nuclear Weapons". India has annually tabled this resolution at the UN General Assembly (UNGA) since 1982. The resolution aims at prohibiting the use or threat of use of nuclear weapons under any circumstances, a step that can substantially reduce the prospect of nuclear use and contribute towards the creation of a climate for a subsequent agreement on the prohibition of nuclear weapons *in toto*.

From the experience of the Russia-Ukraine conflict, it is also clear that an undertaking of negative security assurances by the NWS must simultaneously be accompanied by positive security assurances that commit the NWS to come to the rescue of nation(s) that are threatened with such use. There should be an undertaking that any country using or threatening to use nuclear weapons shall face commensurate retribution and a total boycott by all the countries of the world. This would make nuclear weapons significantly impotent and useless. The value of nuclear weapons would fall and none would want to acquire weapons that could not be used without the fear of serious global reprisals for their action. Therefore, a total ban on the use of nuclear weapons would directly strike at the very root of their utility.

NFU, then, would reduce the likelihood of a nuclear exchange between the NWS. And a ban on the use or threat of use (including negative and positive security assurances) would address the concerns of the NNWS and reduce their temptation to acquire these weapons. Unity of action by the international community can be a major deterrent in modern times when economic interdependence adds to the vulnerabilities of nations.

Nuclear risks pose an existential danger to mankind, much like climate change. Hopefully, the Russia-Ukraine crisis will propel the NWS to seriously introspect on the costs and benefits of retaining nuclear weapons. And, one can only wish that this would come to pass before a nuclear catastrophe of some sort strikes. Mankind cannot hope to live indefinitely with nuclear weapons and remain safe from their use. Sooner rather than later, the weapons could come into use, deliberately or accidentally. And this is a risk not worth taking.

# Index